Fundamentals of Power Electronics

Fundamentals of
Power Electronics

Rick Jacobs

NY RESEARCH
P R E S S

New York

Published by NY Research Press
118-35 Queens Blvd., Suite 400,
Forest Hills, NY 11375, USA
www.nyresearchpress.com

Fundamentals of Power Electronics
Rick Jacobs

International Standard Book Number: 978-1-63238-859-9 (Hardback)

Cataloging-in-Publication Data

Fundamentals of power electronics / Rick Jacobs.
   p. cm.
Includes bibliographical references and index.
ISBN 978-1-63238-859-9
1. Power electronics. 2. Electronics. 3. Electric power. I. Jacobs, Rick.
TK7881.15 .F86 2022
621.317--dc23

Table of Contents

Preface

This book is a culmination of my many years of practice in this field. I attribute the success of this book to my support group. I would like to thank my parents who have showered me with unconditional love and support and my peers and professors for their constant guidance.

The application of solid-state electronics for controlling and converting electric power is referred to as power electronics. Modern power electronics systems use semiconductor switching devices such as diodes, thyristors and power transistors to perform the conversion. The power conversion systems can be classified into various categories such as AC to DC, DC to AC, DC to DC and AC to AC. AC to DC converter is found in many consumer electronic devices such as television sets, battery chargers and personal computers. DC to AC converters are known as power inverters. They are used in numerous devices such as adjustable speed drives, uninterruptible power supplies and flexible AC transmission systems. This book elucidates the concepts and innovative models around prospective developments with respect to modern power electronics. Different approaches, evaluations and principles related to this field have been included herein. Coherent flow of topics, student-friendly language and extensive use of examples make this book an invaluable source of knowledge.

The details of chapters are provided below for a progressive learning:

Chapter – What are Power Electronics?

The application of solid-state electronics to the conversion and control of electric power is referred to as power electronics. It finds its applications in various sectors such as automotives and traction, industries, solar system, defence and aerospace, renewable energy, etc. This chapter has been carefully written to provide an easy understanding of various aspects and applications of power electronics.

Chapter – Important Aspects of Power Electronics

Some of the basic concepts of electronics are voltage, current and EMF. Voltage refers to the difference in electric potential between two points, current is the rate of flow of electric charge and the energy per unit electric charge that is imparted by an energy source is known as EMF or the electromotive force. The topics elaborated in this chapter will help in gaining a better perspective about these basic concepts of electronics.

Chapter – Diode and its Types

A diode refers to a two-terminal electronic component that generally conducts current in one direction. Some of the common types of diodes are zener diode, light emitting diode, Schottky diode, tunnel diode, laser diode, vacuum diode, Gunn diode, photodiode and avalanche diode. These types of diodes have been thoroughly discussed in this chapter.

Chapter – Thyristor: A Semiconductor Device

Thyristor is a solid-state semiconductor device having four layers of alternating p- and n-type materials. DIAC and TRIAC are a few types of thyristors. The diverse applications of these types of thyristors have been thoroughly discussed in this chapter.

Chapter – Rectifiers

The electrical devices which convert alternating current to direct current are known as rectifiers. Common types of rectifiers include single phase uncontrolled rectifiers, single phase controlled rectifiers and three phase rectifiers. The chapter closely examines these types of rectifiers to provide an extensive understanding of the subject.

Chapter – Inverters

A power electronic device that changes direct current into alternating current is known as an inverter. The major types of inverters are single-phase inverter and three-phase inverter. The diverse types of inverters and their applications have been thoroughly discussed in this chapter.

Chapter – Converters

Power electronics converters are used to modify the form of electrical energy. Two major types of converters are DC to DC converter and AC to AC converter. This chapter discusses in detail the various subtypes of these types of converters.

Rick Jacobs

What are Power Electronics?

The application of solid-state electronics to the conversion and control of electric power-er is referred to as power electronics. It finds its applications in various sectors such as automotives and traction, industries, solar system, defence and aerospace, renewable energy, etc. This chapter has been carefully written to provide an easy understanding of various aspects and applications of power electronics.

Power Electronics is the art of converting electrical energy from one form to another in an efficient, clean, compact, and robust manner for convenient utilisation.

A passenger lift in a modern building equipped with a Variable-Voltage-Variable-Speed induction-machine drive offers a comfortable ride and stops exactly at the floor level. Behind the scene it consumes less power with reduced stresses on the motor and cor-ruption of the utility mains.

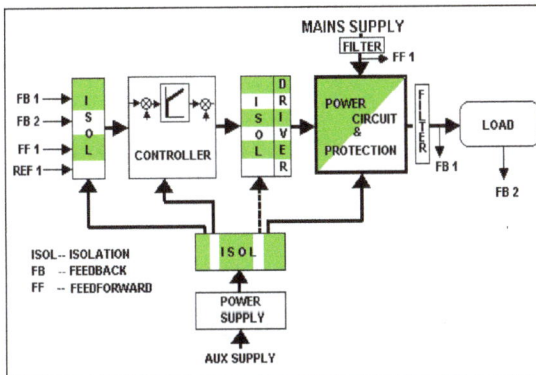

The block diagram of a typical Power Electronic converter.

Power electronics involves the study:

- Power semiconductor devices - their physics, characteristics, drive requirements and their protection for optimum utilisation of their capacities.

- Power converter topologies involving them.

- Control strategies of the converters.

- Digital, analogue and microelectronics involved.

- Capacitive and magnetic energy storage elements.

- Rotating and static electrical devices.

- Quality of waveforms generated.

- Electro Magnetic and Radio Frequency Interference.

- Thermal Management.

Difference between Power Electronics and Linear Electronics

While power management IC's in mobile sets working on Power Electronic principles are meant to handle only a few milliwatts, large linear audio amplifiers are rated at a few thousand watts.

The utilisation of the Bipolar junction transistor, figure in the two types of amplifiers best symbolises the difference. In Power Electronics all devices are operated in the switching mode - either 'FULLY-ON' or 'FULLY-OFF' states. The linear amplifier concentrates on fidelity in signal amplification, requiring transistors to operate strictly in the linear (active) zone, figure saturation and cutoff zones in the V_{CE} - I_C plane are avoided. In a Power electronic switching amplifier, only those areas in the V_{CE} - I_C plane which have been skirted above, are suitable. Onstate dissipation is minimum if the device is in saturation (or quasi-saturation for optimising other losses). In the off-state also, losses are minimum if the BJT is reverse biased. A BJT switch will try to traverse the active zone as fast as possible to minimise switching losses.

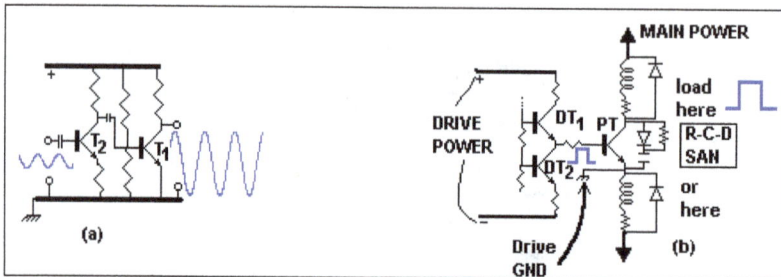

Typical Bipolar transistor based (a) linear (common emitter) (voltage) amplifier stage and (b) switching (power) amplifier.

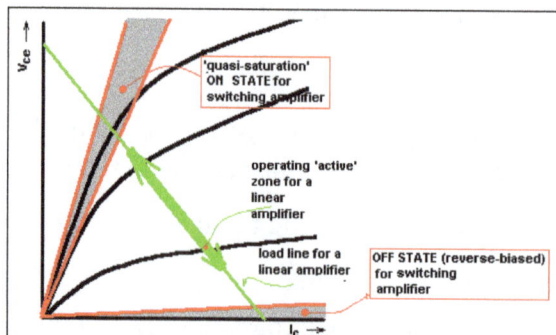

Operating zones for operating a Bipolar Junction Transistor as a linear and a switching amplifier.

Linear Operation	Switching Operation
Active zone selected: Good linearity between input/output.	Active zone avoided : High losses, encountered only during transients.
Saturation & cut-off zones avoided: poor linearity.	Saturation & cut-off (negative bias) zones selected: low losses.
Transistor biased to operate around quiescent point.	No concept of quiescent point.
Common emitter, Common collector, common base modes.	Transistor driven directly at base - emitter and load either on collector or emitter.
Output transistor barely protected.	Switching-Aid-Network (SAN) and other protection to main transistor.
Utilisation of transistor rating of secondary importance.	Utilisation of transistor rating optimised.

An example illustrating the linear and switching solutions to a power supply specification will emphasise the difference.

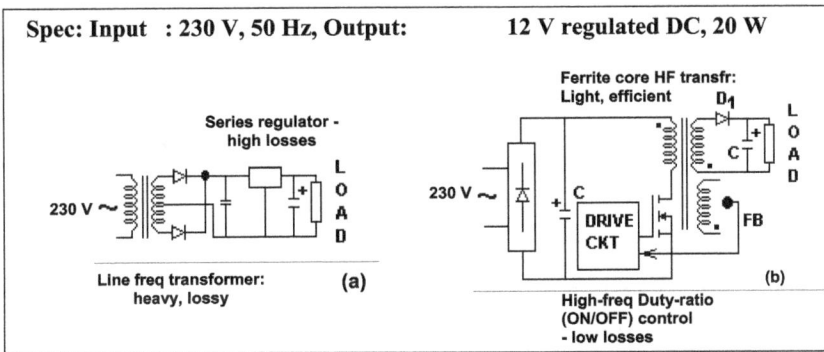

(a) A Linear regulator and (b) a switching regulator solution of the specification above.

The linear solution, figure (a), to this quite common specification would first step down the supply voltage to 12-0-12 V through a power frequency transformer. The output would be rectified using Power frequency diodes, electrolytic capacitor filter and then series regulated using a chip or a audio power transistor. The tantalum capacitor filter would follow. The balance of the voltage between the output of the rectifier and the output drops across the regulator device which also carries the full load current. The power loss is therefore considerable. Also, the stepdown iron-core transformer is both heavy, and lossy. However, only twice-line-frequency ripples appear at the output and material cost and technical know-how required is low.

In the switching solution figure (b) using a MOSFET driven flyback converter, first the line voltage is rectified and then isolated, stepped-down and regulated. A ferrite-core high-frequency (HF) transformer is used. Losses are negligible compared to the first solution and the converter is extremely light. However significant high frequency (related to the switching frequency) noise appear at the output which can only be minimised through the use of costly 'grass' capacitors.

APPLICATIONS OF POWER ELECTRONICS

Power electronics has penetrated almost all the fields where electrical energy is in the picture. This trend is an ever increasing one especially with present trends of new devices and integrated design of power semiconductor devices and controllers. The ease of manufacturing has also led to availability of these devices in a vast range of ratings and gradually has appeared in high voltage and extra high voltage systems also. The day is not far when all of the electrical energy in the world will pass through power electronic systems.

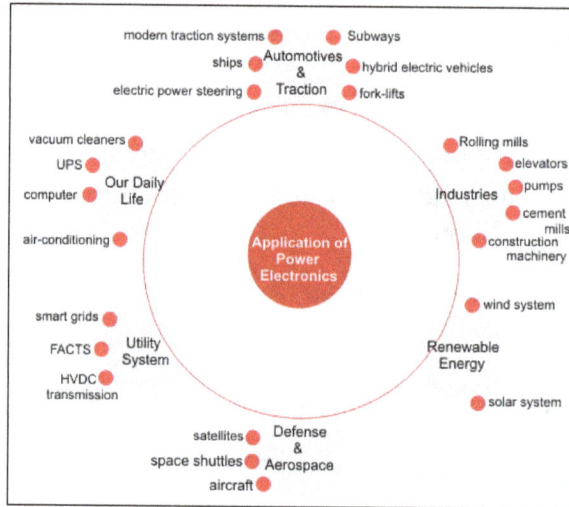

- Our Daily Life: If we look around ourselves, we can find a whole lot of power electronics applications such as a fan regulator, light dimmer, air-conditioning, induction cooking, emergency lights, personal computers, vacuum cleaners, UPS (uninterrupted power system), battery charges, etc.

- Automotives and Traction: Subways, hybrid electric vehicles, trolley, fork-lifts, and many more. A modern car itself has so many components where power electronic is used such as ignition switch, windshield wiper control, adaptive front lighting, interior lighting, electric power steering and so on. Besides power electronics are extensively used in modern traction systems and ships.

- Industries: Almost all the motors employed in the industries are controlled by power electronic drives, for eg. Rolling mills, textile mills, cement mills, compressors, pumps, fans, blowers, elevators, rotary kilns etc. Other applications include welding, arc furnace, cranes, heating applications, emergency power systems, construction machinery, excavators etc.

- Defense and Aerospace: Power supplies in aircraft, satellites, space shuttles, advance control in missiles, unmanned vehicles and other defense equipments.

- Renewable Energy: Generation systems such as solar, wind etc. needs power conditioning systems, storage systems and conversion systems in order to become usable. For example solar cells generate DC power and for general application we need AC power and hence power electronic converter is used.

- Utility System: HVDC transmission, VAR compensation (SVC), static circuit breakers, generator excitation systems, FACTS, smart grids, etc.

Application of Power Electronics in Automotive Applications

Application of power electronics in automotive applications plays a major role in controlling automotive electronics. Automotive electronics include modern electric power steering, HEV main inverter, central body control, braking system, seat control, and so on.

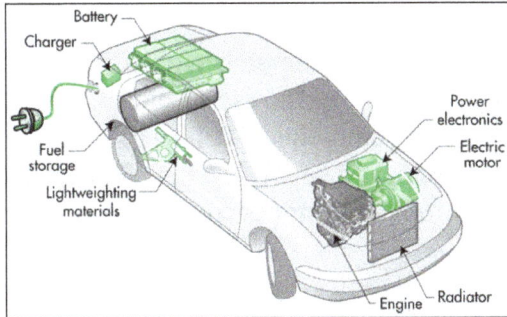

Power Electronics in Automotive Applications.

In our day-to-day life, we frequently observe heat radiating from car engine after the car has been driven for a certain distance. This is due to power train system of automotive electronics with an engine or internal combustion or motor as one of the subsystem operating with high temperature exceeding 125 degrees Celsius. Application of power electronics with components such as silicon-based power MOSFETs and IGBTs that are used as power electronic switches in the power train system of automotive electrical and electronic systems for reducing the overall size. And also for managing thermal issues in which a high power of kW range is being used for improving fuel efficiency.

Silicon based Dual Channel MOSFET.

Limitations can be overcome by using a wideband gap semiconductors like silicon carbide with a high-operating temperature that allows placing the circuit near high temperature location. It has two or three times higher thermal conductivity than silicon, which will eliminate need of big copper blocks and water jackets. Silicon carbide has high breakdown voltage and capable of switching at high frequencies with very less power loss which makes the overall size of circuitry very small.

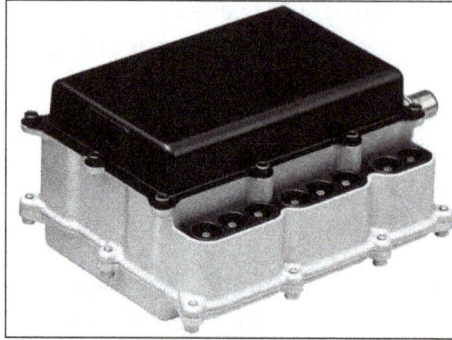

Silicon carbide chip.

Power electronics applications are extended to various fields such as Aerospace, Automotive electrical and electronic systems, commercial, industrial, residential, telecommunication, transportation, utility systems, etc. In case of automotive electronics, the electrically-generated systems are used in automobiles such as road vehicles like telematics, in-car entertainment systems, carputers, and so on. The need to control engines of automobiles originated in automotive electronics for proper controlling and conversion.

Automotive electronics components.

Automotive electronics are classified into different types: engine electronics, transmission electronics, chassis electronics, active safety, driver assistance, passenger comfort and entertainment systems. For any power system such as DC/DC or DC/AC or AC/DC, the power electronic components like controllers, gate drivers, converters and so on are required. Generally, based on the vehicle or power supply manufacturer requirements the analog or digital controllers are chosen such that the following parameters including cost, integration, reliability and flexibility are taken into consideration.

Power Electronics Application in Automotive Electronics

Applications of power electronics in automotive electrical and electronic systems includes high voltage systems, automotive power generation, switched mode power supply (SMPS), DC to DC converters, electric drives, traction inverter or DC to AC converter, power electronic component, high temperature requirement, application of SMPS in power train system, and so on. For example, consider a modern car, in which we can find many power electronic components such as ignition switch, control module, vehicle speed sensor, steering sensor and other components, as shown in figure.

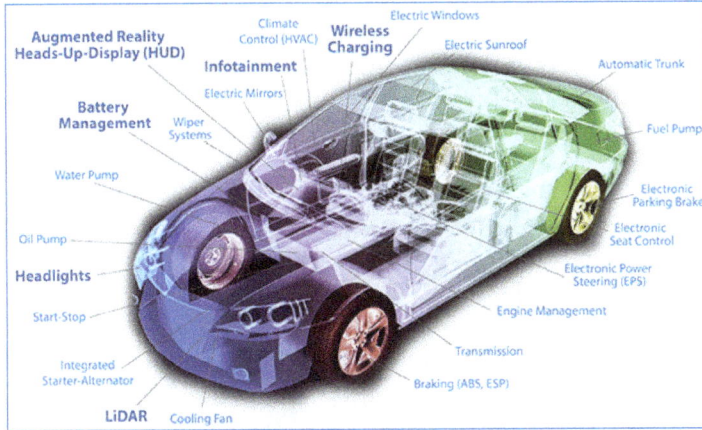

Power electronics application in automotive electronics.

Automotive Power Generation

Application of power electronics in the automotive power generation system provides automotive alternators with improved efficiency and high power, along with high temperature withstanding capacity and high-power density with a variety of research in designing of alternator with a switched mode power electronics applications. The frequently used alternator in automotive applications is Lundell or Claw-pole alternator, as it is suitable for the required emerging performance. Field and armature characteristics of this alternator are enhanced by the use of power electronics. These alternators are used in automobiles for supplying power to the batteries and electrical system while the engine is running. Automotive alternators require a power electronic voltage regulator for producing a constant voltage at the battery terminals by modulating small field current.

Cut view of Lundell Alternator.

Switched Mode Power Supply (SMPS)

SMPS concept is based on the power electronics devices such as semiconductor devices that operates in an on state that has zero voltage and an off state that has zero current during this state theoretically with 100% efficiency. To switch these power semiconductor devices on and off the pulse width modulation (PWM) technique is used. Less bulky and small-sized power electronics based converters are used for high frequency switching as these switches are capable of operating under high switching frequencies.

SMPS.

SMPS Applications in the Power Train System

The power train systems of HEVs, electric vehicles and ICE need the following SMPS conditioners such as:

- Regenerative braking (AC/DC),
- On-board charger (AC/DC),
- Dual-battery system (DC/DC),
- Traction motor (DC/AC).

DC to DC Converters

There are different DC to DC converter topologies available which can be used based on the requirements. These topologies are classified as isolated and non-isolated topologies which are adopted in power train systems. The application of power electronics in switching has brought a concept of soft-switching where the switches are subjected to low stress using an LLC or resonant mode. These soft-switching, highly reliable and longlife converters are very useful in the automotive electronics market. There are bi-directional converters such as 400 to 12V for electric vehicles and 48 to 12V for hybrid electric vehicle or internal combustion engine.

DC-DC Converter.

Traction Inverter (DC/AC)

Electrical motors are machines used for converting electrical energy into mechanical energy and primarily DC motors are used for this purpose, but due to the unreliability of DC motors, AC motors are used because of their efficiency. Application of power electronics in building controllers for AC motors has tremendous progress from the past two decades. Thus, for AC motors to supply power, power stored in batteries of the automotive electrical and electronic systems of electric vehicles or hybrid electric vehicles or ICE require the application of power electronics such as DC to AC converters or electrical inverters.

SPI Inverter.

On-board Charger (AC/DC)

Vehicles with automotive electronics consist of batteries that need to be charged; for this charging purpose, the supply AC power has to be converted into DC. We know that, the power can be stored in batteries only in the form of DC. This conversion of AC to DC can be done by the application of power electronics converters called as rectifiers.

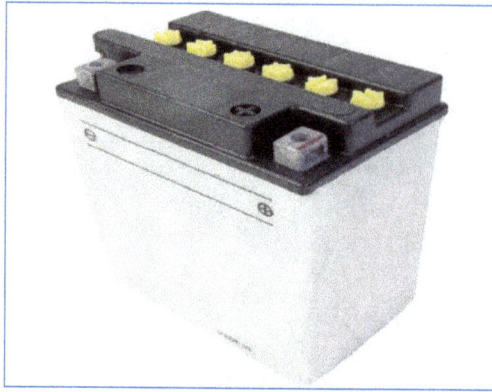

Automotive Batteries.

The application of power electronics is increasing with the advancing technologies in automotive electrical and electronics systems for improving the overall system efficiency with high operating temperature, increasing flexibility, reliability and to reduce the overall size of the circuitry.

2

Important Aspects of Power Electronics

Some of the basic concepts of electronics are voltage, current and EMF. Voltage refers to the difference in electric potential between two points, current is the rate of flow of electric charge and the energy per unit electric charge that is imparted by an energy source is known as EMF or the electromotive force. The topics elaborated in this chapter will help in gaining a better perspective about these basic concepts of electronics.

VOLTAGE

Voltage is a quantitative expression of the potential difference in charge between two points in an electrical field.

The greater the voltage, the greater the flow of electrical current (that is, the quantity of charge carriers that pass a fixed point per unit of time) through a conducting or semiconducting medium for a given resistance to the flow. Voltage is symbolized by an uppercase italic letter V or E. The standard unit is the volt, symbolized by a non-italic uppercase letter V. One volt will drive one coulomb (6.24×10^{18}) charge carriers, such as electrons, through a resistance of one ohm in one second.

Voltage can be direct or alternating. A direct voltage maintains the same polarity at all times. In an alternating voltage, the polarity reverses direction periodically. The number of complete cycles per second is the frequency, which is measured in hertz (one cycle per second), kilohertz, megahertz, gigahertz, or terahertz. An example of direct voltage is the potential difference between the terminals of an electrochemical cell. Alternating voltage exists between the terminals of a common utility outlet.

A voltage produces an electrostatic field, even if no charge carriers move (that is, no current flows). As the voltage increases between two points separated by a specific distance, the electrostatic field becomes more intense. As the separation increases between two points having a given voltage with respect to each other, the electrostatic flux density diminishes in the region between them.

Volt

The volt (symbol: V) is the derived unit for electric potential, electric potential difference, and electromotive force. The volt is named in honour of the Italian physicist Alessandro Volta, who invented the voltaic pile, possibly the first chemical battery.

Hydraulic Analogy

A simple analogy for an electric circuit is water flowing in a closed circuit of pipework, driven by a mechanical pump. This can be called a "water circuit". Potential difference between two points corresponds to the pressure difference between two points. If the pump creates a pressure difference between two points, then water flowing from one point to the other will be able to do work, such as driving a turbine. Similarly, work can be done by an electric current driven by the potential difference provided by a battery. For example, the voltage provided by a sufficiently-charged automobile battery can "push" a large current through the windings of an automobile's starter motor. If the pump isn't working, it produces no pressure difference, and the turbine will not rotate. Likewise, if the automobile's battery is very weak or "dead" (or "flat"), then it will not turn the starter motor.

The hydraulic analogy is a useful way of understanding many electrical concepts. In such a system, the work done to move water is equal to the pressure multiplied by the volume of water moved. Similarly, in an electrical circuit, the work done to move electrons or other charge-carriers is equal to "electrical pressure" multiplied by the quantity of electrical charges moved. In relation to "flow", the larger the "pressure difference" between two points (potential difference or water pressure difference), the greater the flow between them (electric current or water flow).

Applications

Working on high voltage power lines.

Specifying a voltage measurement requires explicit or implicit specification of the points across which the voltage is measured. When using a voltmeter to measure potential difference, one electrical lead of the voltmeter must be connected to the first point, one to the second point.

A common use of the term "voltage" is in describing the voltage dropped across an electrical device (such as a resistor). The voltage drop across the device can be understood as the difference between measurements at each terminal of the device with respect to a common reference point (or ground). The voltage drop is the difference between the two readings. Two points in an electric circuit that are connected by an ideal conductor without resistance and not within a changing magnetic field have a voltage of zero. Any two points with the same potential may be connected by a conductor and no current will flow between them.

Addition of Voltages

The voltage between A and C is the sum of the voltage between A and B and the voltage between B and C. The various voltages in a circuit can be computed using Kirchhoff's circuit laws.

When talking about alternating current (AC) there is a difference between instantaneous voltage and average voltage. Instantaneous voltages can be added for direct current (DC) and AC, but average voltages can be meaningfully added only when they apply to signals that all have the same frequency and phase.

Measuring Instruments

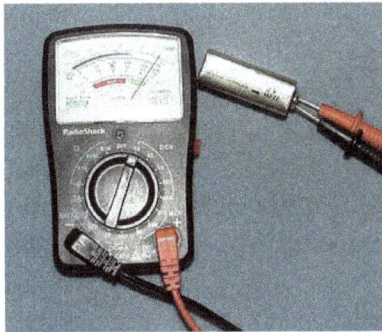

Multimeter set to measure voltage.

Instruments for measuring voltages include the voltmeter, the potentiometer, and the oscilloscope. Analog voltmeters, such as moving-coil instruments, work by measuring the current through a fixed resistor, which, according to Ohm's Law, is proportional to the voltage across the resistor. The potentiometer works by balancing the unknown voltage against a known voltage in a bridge circuit. The cathode-ray oscilloscope works by amplifying the voltage and using it to deflect an electron beam from a straight path, so that the deflection of the beam is proportional to the voltage.

Typical Voltages

A common voltage for flashlight batteries is 1.5 volts (DC). A common voltage for automobile batteries is 12 volts (DC).

Common voltages supplied by power companies to consumers are 110 to 120 volts (AC) and 220 to 240 volts (AC). The voltage in electric power transmission lines used to distribute electricity from power stations can be several hundred times greater than consumer voltages, typically 110 to 1200 kV (AC).

The voltage used in overhead lines to power railway locomotives is between 12 kV and 50 kV (AC) or between 1.5 kV and 3 kV (DC).

Galvani Potential vs. Electrochemical Potential

Inside a conductive material, the energy of an electron is affected not only by the average electric potential, but also by the specific thermal and atomic environment that it is in. When a voltmeter is connected between two different types of metal, it measures not the electrostatic potential difference, but instead something else that is affected by thermodynamics. The quantity measured by a voltmeter is the negative of the difference of the electrochemical potential of electrons (Fermi level) divided by the electron charge and commonly referred to as the voltage difference, while the pure unadjusted electrostatic potential (not measurable with a voltmeter) is sometimes called Galvani potential. The terms "voltage" and "electric potential" are ambiguous in that, in practice, they can refer to *either* of these in different contexts.

CURRENT

An electric current is the rate of flow of electric charge past a point or region. An electric current is said to exist when there is a net flow of electric charge through a region. In electric circuits this charge is often carried by electrons moving through a wire. It can also be carried by ions in an electrolyte, or by both ions and electrons such as in an ionized gas (plasma).

The SI unit of electric current is the ampere, which is the flow of electric charge across a surface at the rate of one coulomb per second. The ampere (symbol: A) is an SI base unit Electric current is measured using a device called an ammeter.

Electric currents cause Joule heating, which creates light in incandescent light bulbs. They also create magnetic fields, which are used in motors, inductors and generators.

The moving charged particles in an electric current are called charge carriers. In metals, one or more electrons from each atom are loosely bound to the atom, and can move freely about within the metal. These conduction electrons are the charge carriers in metal conductors.

Symbol

The conventional symbol for current is I, which originates from the French phrase *intensité du courant*, (current intensity). Current intensity is often referred to simply as *current*. The *I* symbol was used by André-Marie Ampère, after whom the unit of electric current is named, in formulating Ampère's force law (1820). The notation travelled from France to Great Britain, where it became standard, although at least one journal did not change from using *C* to *I* until 1896.

Conventions

In a conductive material, the moving charged particles that constitute the electric current are called charge carriers. In metals, which make up the wires and other conductors in most electrical circuits, the positively charged atomic nuclei of the atoms are held in a fixed position, and the negatively charged electrons are the charge carriers, free to move about in the metal. In other materials, notably the semiconductors, the charge carriers can be positive *or* negative, depending on the dopant used. Positive and negative charge carriers may even be present at the same time, as happens in an electrolyte in an electrochemical cell.

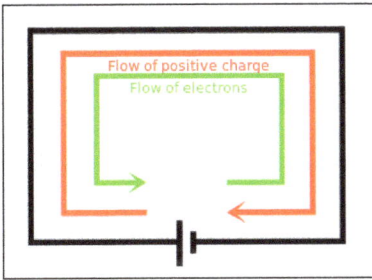

The electrons, the charge carriers in an electrical circuit, flow in the opposite direction of the conventional electric current.

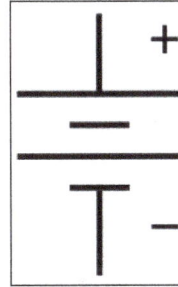

The symbol for a battery in a circuit diagram.

A flow of positive charges gives the same electric current, and has the same effect in a circuit, as an equal flow of negative charges in the opposite direction. Since current can be the flow of either positive or negative charges, or both, a convention is needed for the direction of current that is independent of the type of charge carriers. The direction of *conventional current* is arbitrarily defined as the same direction as positive charges flow.

Since electrons, the charge carriers in metal wires and most other parts of electric circuits, have a negative charge, as a consequence, they flow in the opposite direction of conventional current flow in an electrical circuit.

Reference Direction

Since the current in a wire or component can flow in either direction, when a variable *I* is defined to represent that current, the direction representing positive current must be specified, usually by an arrow on the circuit schematic diagram. This is called the *reference direction* of current *I*. If the current flows in the opposite direction, the variable *I* has a negative value.

When analyzing electrical circuits, the actual direction of current through a specific circuit element is usually unknown. Consequently, the reference directions of currents are often assigned arbitrarily. When the circuit is solved, a negative value for the variable means that the actual direction of current through that circuit element is opposite that of the chosen reference direction. In electronic circuits, the reference current directions are often chosen so that all currents are toward ground. This often corresponds to the actual current direction, because in many circuits the power supply voltage is positive with respect to ground.

Ohm's Law

Ohm's law states that the current through a conductor between two points is directly proportional to the potential difference across the two points. Introducing the constant of proportionality, the resistance, one arrives at the usual mathematical equation that describes this relationship:

$$I = \frac{V}{R}$$

where *I* is the current through the conductor in units of amperes, *V* is the potential difference measured *across* the conductor in units of volts, and *R* is the resistance of the conductor in units of ohms. More specifically, Ohm's law states that the *R* in this relation is constant, independent of the current.

Alternating and Direct Current

In alternating current (AC) systems, the movement of electric charge periodically reverses direction. AC is the form of electric power most commonly delivered to businesses and residences. The usual waveform of an AC power circuit is a sine wave. Certain applications use different waveforms, such as triangular or square waves. Audio and radio signals carried on electrical wires are also examples of alternating current. An important goal in these applications is recovery of information encoded (or *modulated*) onto the AC signal.

In contrast, direct current (DC) is the unidirectional flow of electric charge, or a system in which the movement of electric charge is in one direction only. Direct current is produced by sources such as batteries, thermocouples, solar cells, and commutator-type electric machines of the dynamo type. Direct current may flow in a conductor such as a wire, but can also flow through semiconductors, insulators, or even through a vacuum as in electron or ion beams. An old name for direct current was *galvanic current*.

Occurrences

Natural observable examples of electrical current include lightning, static electric discharge, and the solar wind, the source of the polar auroras.

Man-made occurrences of electric current include the flow of conduction electrons in metal wires such as the overhead power lines that deliver electrical energy across long distances and the smaller wires within electrical and electronic equipment. Eddy currents are electric currents that occur in conductors exposed to changing magnetic fields. Similarly, electric currents occur, particularly in the surface, of conductors exposed to electromagnetic waves. When oscillating electric currents flow at the correct voltages within radio antennas, radio waves are generated.

In electronics, other forms of electric current include the flow of electrons through resistors or through the vacuum in a vacuum tube, the flow of ions inside a battery or a neuron, and the flow of holes within metals and semiconductors.

Current Measurement

Current can be measured using an ammeter.

Electric current can be directly measured with a galvanometer, but this method involves breaking the electrical circuit, which is sometimes inconvenient.

Current can also be measured without breaking the circuit by detecting the magnetic field associated with the current. Devices, at the circuit level, use various techniques to measure current:

- Shunt resistors,

- Hall effect current sensor transducers,

- Transformers (however DC cannot be measured),

- Magnetoresistive field sensors,

- Rogowski coils,

- Current clamps.

Resistive Heating

Joule heating, also known as *ohmic heating* and *resistive heating*, is the process of power dissipation by which the passage of an electric current through a conductor increases the internal energy of the conductor, converting thermodynamic work into heat. The phenomenon was first studied by James Prescott Joule in 1841. Joule immersed a length of wire in a fixed mass of water and measured the temperature rise due to a known current through the wire for a 30 minute period. By varying the current and the length of the wire he deduced that the heat produced was proportional to the square of the current multiplied by the electrical resistance of the wire:

$$P \propto I^2 R$$

This relationship is known as Joule's Law. The SI unit of energy was subsequently named the joule and given the symbol J. The commonly known SI unit of power, the watt (symbol: W), is equivalent to one joule per second.

Electromagnetism

Electromagnet

In an electromagnet a coil of wires behaves like a magnet when an electric current flows through it. When the current is switched off, the coil loses its magnetism immediately.

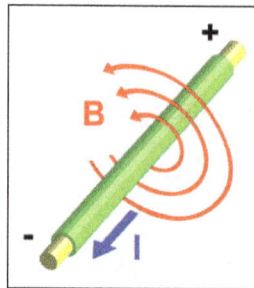

According to Ampère's circuital law, an electric current produces a magnetic field.

Electric current produces a magnetic field. The magnetic field can be visualized as a pattern of circular field lines surrounding the wire that persists as long as there is current.

Electromagnetic Induction

Magnetic fields can also be used to make electric currents. When a changing magnetic field is applied to a conductor, an electromotive force (EMF) is induced, which starts an electric current, when there is a suitable path.

Radio Waves

When an electric current flows in a suitably shaped conductor at radio frequencies, radio waves can be generated. These travel at the speed of light and can cause electric currents in distant conductors.

Conduction Mechanisms in Various Media

In metallic solids, electric charge flows by means of electrons, from lower to higher electrical potential. In other media, any stream of charged objects (ions, for example) may constitute an electric current. To provide a definition of current independent of the type of charge carriers, *conventional current* is defined as moving in the same direction as the positive charge flow. So, in metals where the charge carriers (electrons) are negative, conventional current is in the opposite direction as the electrons. In conductors where the charge carriers are positive, conventional current is in the same direction as the charge carriers.

In a vacuum, a beam of ions or electrons may be formed. In other conductive materials, the electric current is due to the flow of both positively and negatively charged particles at the same time. In still others, the current is entirely due to positive charge flow. For example, the electric currents in electrolytes are flows of positively and negatively charged ions. In a common lead-acid electrochemical cell, electric currents are composed of positive hydronium ions flowing in one direction, and negative sulfate ions flowing in the other. Electric currents in sparks or plasma are flows of electrons as well as positive and negative ions. In ice and in certain solid electrolytes, the electric current is entirely composed of flowing ions.

Metals

In a metal, some of the outer electrons in each atom are not bound to the individual atom as they are in insulating materials, but are free to move within the metal lattice. These conduction electrons can serve as charge carriers, carrying a current. Metals are particularly conductive because there are a large number of these free electrons, typically one per atom in the lattice. With no external electric field applied, these electrons move about randomly due to thermal energy but, on average, there is zero net current within the metal. At room temperature, the average speed of these random motions is 10^6 metres per second. Given a surface through which a metal wire passes, electrons move in both directions across the surface at an equal rate.

When a metal wire is connected across the two terminals of a DC voltage source such as a battery, the source places an electric field across the conductor. The moment contact is made, the free electrons of the conductor are forced to drift toward the positive terminal under the influence of this field. The free electrons are therefore the charge carrier in a typical solid conductor.

For a steady flow of charge through a surface, the current I (in amperes) can be calculated with the following equation:

$$I = \frac{Q}{t},$$

where Q is the electric charge transferred through the surface over a time t. If Q and t are measured in coulombs and seconds respectively, I is in amperes.

More generally, electric current can be represented as the rate at which charge flows through a given surface as:

$$I = \frac{dQ}{dt}.$$

Electrolytes

Electric currents in electrolytes are flows of electrically charged particles (ions). For example, if an electric field is placed across a solution of Na^+ and Cl^- (and conditions are right) the sodium ions move towards the negative electrode (cathode), while the chloride ions move towards the positive electrode (anode). Reactions take place at both electrode surfaces, neutralizing each ion.

Water-ice and certain solid electrolytes called proton conductors contain positive hydrogen ions ("protons") that are mobile. In these materials, electric currents are composed of moving protons, as opposed to the moving electrons in metals.

In certain electrolyte mixtures, brightly coloured ions are the moving electric charges. The slow progress of the colour makes the current visible.

Gases and Plasmas

In air and other ordinary gases below the breakdown field, the dominant source of electrical conduction is via relatively few mobile ions produced by radioactive gases, ultraviolet light, or cosmic rays. Since the electrical conductivity is low, gases are dielectrics or insulators. However, once the applied electric field approaches the breakdown value, free electrons become sufficiently accelerated by the electric field to create additional free electrons by colliding, and ionizing, neutral gas atoms or molecules in a process called avalanche breakdown. The breakdown process forms a plasma that contains enough mobile electrons and positive ions to make it an electrical conductor. In the process, it forms a light emitting conductive path, such as a spark, arc or lightning.

Plasma is the state of matter where some of the electrons in a gas are stripped or "ionized" from their molecules or atoms. A plasma can be formed by high temperature, or by application of a high electric or alternating magnetic field as noted. Due to their lower mass, the electrons in a plasma accelerate more quickly in response to an electric field than the heavier positive ions, and hence carry the bulk of the current. The free ions recombine to create new chemical compounds (for example, breaking atmospheric oxygen into single oxygen [$O_2 \rightarrow 2O$], which then recombine creating ozone [O_3]).

Vacuum

Since a "perfect vacuum" contains no charged particles, it normally behaves as a perfect insulator. However, metal electrode surfaces can cause a region of the vacuum to become conductive by injecting free electrons or ions through either field electron emission or thermionic emission. Thermionic emission occurs when the thermal energy exceeds the metal's work function, while field electron emission occurs when the electric field at the surface of the metal is high enough to cause tunneling, which results in the ejection of free electrons from the metal into the vacuum. Externally heated electrodes are often used to generate an electron cloud as in the filament or indirectly heated cathode of vacuum tubes. Cold electrodes can also spontaneously produce electron clouds via thermionic emission when small incandescent regions (called cathode spots or anode spots) are formed. These are incandescent regions of the electrode surface that are created by a localized high current. These regions may be initiated by field electron emission, but are then sustained by localized thermionic emission once a vacuum arc forms. These small electron-emitting regions can form quite rapidly, even explosively, on a metal surface subjected to a high electrical field. Vacuum tubes and sprytrons are some of the electronic switching and amplifying devices based on vacuum conductivity.

Superconductivity

Superconductivity is a phenomenon of exactly zero electrical resistance and expulsion of magnetic fields occurring in certain materials when cooled below a characteristic critical temperature. It was discovered by Heike Kamerlingh Onnes on April 8, 1911 in Leiden. Like ferromagnetism and atomic spectral lines, superconductivity is a quantum mechanical phenomenon. It is characterized by the Meissner effect, the complete ejection of magnetic field lines from the interior of the superconductor as it transitions into the superconducting state. The occurrence of the Meissner effect indicates that superconductivity cannot be understood simply as the idealization of *perfect conductivity* in classical physics.

Semiconductor

In a semiconductor it is sometimes useful to think of the current as due to the flow of positive "holes" (the mobile positive charge carriers that are places where the semiconductor crystal is missing a valence electron). This is the case in a p-type semiconductor. A semiconductor has electrical conductivity intermediate in magnitude between that of a conductor and an insulator. This means a conductivity roughly in the range of 10^{-2} to 10^4 siemens per centimeter (S cm^{-1}).

In the classic crystalline semiconductors, electrons can have energies only within certain bands (i.e. ranges of levels of energy). Energetically, these bands are located between the energy of the ground state, the state in which electrons are tightly bound to the atomic nuclei of the material, and the free electron energy, the latter describing the energy required for an electron to escape entirely from the material. The energy bands each correspond to a large number of discrete quantum states of the electrons, and most of the states with low energy (closer to the nucleus) are occupied, up to a particular band called the *valence band*. Semiconductors and insulators are distinguished from metals because the valence band in any given metal is nearly filled with electrons under usual operating conditions, while very few (semiconductor) or virtually none (insulator) of them are available in the *conduction band*, the band immediately above the valence band.

The ease of exciting electrons in the semiconductor from the valence band to the conduction band depends on the band gap between the bands. The size of this energy band gap serves as an arbitrary dividing line (roughly 4 eV) between semiconductors and insulators.

With covalent bonds, an electron moves by hopping to a neighboring bond. The Pauli exclusion principle requires that the electron be lifted into the higher anti-bonding state of that bond. For delocalized states, for example in one dimension – that is in a nanowire, for every energy there is a state with electrons flowing in one direction and another state with the electrons flowing in the other. For a net current to flow, more states for one direction than for the other direction must be occupied. For this to occur, energy is required, as in the semiconductor the next higher states lie above the band gap. Often this is stated as: full bands do not contribute to the electrical conductivity. However, as a semiconductor's temperature rises above absolute zero, there is more energy in the semiconductor to spend on lattice vibration and on exciting electrons into the conduction band. The current-carrying electrons in the conduction band are known as *free electrons*, though they are often simply called *electrons* if that is clear in context.

Current Density and Ohm's Law

Current density is the rate at which charge passes through a chosen unit area. It is defined as a vector whose magnitude is the current per unit cross-sectional area. As discussed in Reference direction, the direction is arbitrary. Conventionally, if the moving charges are positive, then the current density has the same sign as the velocity of the charges. For negative charges, the sign of the current density is opposite to the velocity of the charges. In SI units, current density (symbol: j) is expressed in the SI base units of amperes per square metre.

In linear materials such as metals, and under low frequencies, the current density across the conductor surface is uniform. In such conditions, Ohm's law states that the current is directly proportional to the potential difference between two ends (across) of that metal (ideal) resistor (or other ohmic device):

$$I = \frac{V}{R},$$

where I is the current, measured in amperes; V is the potential difference, measured in volts; and R is the resistance, measured in ohms. For alternating currents, especially at higher frequencies, skin effect causes the current to spread unevenly across the conductor cross-section, with higher density near the surface, thus increasing the apparent resistance.

Drift Speed

The mobile charged particles within a conductor move constantly in random directions, like the particles of a gas. (More accurately, a Fermi gas.) To create a net flow of charge, the particles must also move together with an average drift rate. Electrons are the charge carriers in most metals and they follow an erratic path, bouncing from atom to atom, but generally drifting in the opposite direction of the electric field. The speed they drift at can be calculated from the equation:

$$I = nAvQ,$$

where,

> I is the electric current,
>
> n is number of charged particles per unit volume (or charge carrier density),
>
> A is the cross-sectional area of the conductor,
>
> v is the drift velocity,
>
> Q is the charge on each particle.

Typically, electric charges in solids flow slowly. For example, in a copper wire of cross-section 0.5 mm², carrying a current of 5 A, the drift velocity of the electrons is on the order of a millimetre per second. To take a different example, in the near-vacuum inside a cathode ray tube, the electrons travel in near-straight lines at about a tenth of the speed of light.

Any accelerating electric charge, and therefore any changing electric current, gives rise to an electromagnetic wave that propagates at very high speed outside the surface of the conductor. This speed is usually a significant fraction of the speed of light, as can be deduced from Maxwell's Equations, and is therefore many times faster than the drift velocity of the electrons. For example, in AC power lines, the waves of electromagnetic energy propagate through the space between the wires, moving from a source to a distant load, even though the electrons in the wires only move back and forth over a tiny distance.

The ratio of the speed of the electromagnetic wave to the speed of light in free space is called the velocity factor, and depends on the electromagnetic properties of the conductor and the insulating materials surrounding it, and on their shape and size.

The magnitudes (not the natures) of these three velocities can be illustrated by an analogy with the three similar velocities associated with gases:

- The low drift velocity of charge carriers is analogous to air motion; in other words, winds.

- The high speed of electromagnetic waves is roughly analogous to the speed of sound in a gas (sound waves move through air much faster than large-scale motions such as convection).

- The random motion of charges is analogous to heat – the thermal velocity of randomly vibrating gas particles.

EMF

Electromotive force, abbreviation E or emf is the energy per unit electric charge that is imparted by an energy source, such as an electric generator or a battery. Energy is converted from one form to another in the generator or battery as the device does work on the electric charge being transferred within itself. One terminal of the device becomes positively charged, the other becomes negatively charged. The work done on a unit of electric charge, or the energy thereby gained per unit electric charge, is the electromotive force. Electromotive force is the characteristic of any energy source capable of driving electric charge around a circuit. It is abbreviated E in the international metric system but also, popularly, as emf.

Despite its name, electromotive force is not actually a force. It is commonly measured in units of volts, equivalent in the metre–kilogram–second system to one joule per coulomb of electric charge. In the electrostatic units of the centimetre–gram–second system, the unit of electromotive force is the statvolt, or one erg per electrostatic unit of charge.

References

- Bagotskii, Vladimir Sergeevich (2006). Fundamentals of electrochemistry. p. 22. ISBN 978-0-471-70058-6

- Voltage, definition: techtarget.com, Retrieved 7 January, 2019

- C. J. Brockman, "The origin of voltaic electricity: The contact vs. Chemical theory before the concept of E. M. F. Was developed", Journal of Chemical Education, vol. 5, no. 5, pp. 549–555, May 1928

- Electromotive-force, science: britannica.com, Retrieved 8 February, 2019

- Walker, Jearl; Halliday, David; Resnick, Robert (2014). Fundamentals of physics (10th ed.). Hoboken, NJ: Wiley. ISBN 9781118230732. OCLC 950235056

3

Diode and its Types

A diode refers to a two-terminal electronic component that generally conducts current in one direction. Some of the common types of diodes are zener diode, light emitting diode, Schottky diode, tunnel diode, laser diode, vacuum diode, Gunn diode, photodiode and avalanche diode. These types of diodes have been thoroughly discussed in this chapter.

A diode is a specialized electronic component with two electrodes called the anode and the cathode. Most diodes are made with semiconductor materials such as silicon, germanium, or selenium. Some diodes are comprised of metal electrodes in a chamber evacuated or filled with a pure elemental gas at low pressure. Diodes can be used as rectifiers, signal limiters, voltage regulators, switches, signal modulators, signal mixers, signal demodulators, and oscillators.

The fundamental property of a diode is its tendency to conduct electric current in only one direction. When the cathode is negatively charged relative to the anode at a voltage greater than a certain minimum called forward breakover, then current flows through the diode. If the cathode is positive with respect to the anode, is at the same voltage as the anode, or is negative by an amount less than the forward breakover voltage, then the diode does not conduct current. This is a simplistic view, but is true for diodes operating as rectifiers, switches, and limiters. The forward breakover voltage is approximately six tenths of a volt (0.6 V) for silicon devices, 0.3 V for germanium devices, and 1 V for selenium devices.

The above general rule notwithstanding, if the cathode voltage is positive relative to the anode voltage by a great enough amount, the diode will conduct current. The voltage required to produce this phenomenon, known as the avalanche voltage, varies greatly depending on the nature of the semiconductor material from which the device is fabricated. The avalanche voltage can range from a few volts up to several hundred volts.

When an analog signal passes through a diode operating at or near its forward breakover point, the signal waveform is distorted. This nonlinearity allows for modulation, demodulation, and signal mixing. In addition, signals are generated at harmonics, or integral multiples of the input frequency. Some diodes also have a characteristic that is imprecisely termed negative resistance. Diodes of this type, with the application of a voltage at the correct level and the polarity, generate analog signals at microwave radio frequencies.

Semiconductor diodes can be designed to produce direct current (DC) when visible light, infrared transmission (IR), or ultraviolet (UV) energy strikes them. These diodes are known as photovoltaic cells and are the basis for solar electric energy systems and photosensors. Yet another form of diode, commonly used in electronic and computer equipment, emits visible light or IR energy when current passes through it. Such a device is the familiar light-emitting diode (LED).

Main Functions

The most common function of a diode is to allow an electric current to pass in one direction (called the diode's forward direction), while blocking it in the opposite direction (the reverse direction). As such, the diode can be viewed as an electronic version of a check valve. This unidirectional behavior is called rectification, and is used to convert alternating current (ac) to direct current (dc). Forms of rectifiers, diodes can be used for such tasks as extracting modulation from radio signals in radio receivers.

However, diodes can have more complicated behavior than this simple on–off action, because of their nonlinear current-voltage characteristics. Semiconductor diodes begin conducting electricity only if a certain threshold voltage or cut-in voltage is present in the forward direction (a state in which the diode is said to be forward-biased). The voltage drop across a forward-biased diode varies only a little with the current, and is a function of temperature; this effect can be used as a temperature sensor or as a voltage reference. Also, diodes' high resistance to current flowing in the reverse direction suddenly drops to a low resistance when the reverse voltage across the diode reaches a value called the breakdown voltage.

A semiconductor diode's current–voltage characteristic can be tailored by selecting the semiconductor materials and the doping impurities introduced into the materials during manufacture. These techniques are used to create special-purpose diodes that perform many different functions. For example, diodes are used to regulate voltage (Zener diodes), to protect circuits from high voltage surges (avalanche diodes), to electronically tune radio and TV receivers (varactor diodes), to generate radio-frequency oscillations (tunnel diodes, Gunn diodes, IMPATT diodes), and to produce light (light-emitting diodes). Tunnel, Gunn and IMPATT diodes exhibit negative resistance, which is useful in microwave and switching circuits.

Diodes, both vacuum and semiconductor, can be used as shot-noise generators.

Vacuum Tube Diodes

A thermionic diode is a thermionic-valve device consisting of a sealed, evacuated glass or metal envelope containing two electrodes: a cathode and a plate. The cathode is either *indirectly heated* or *directly heated*. If indirect heating is employed, a heater is included in the envelope.

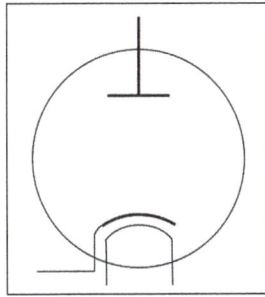

The symbol for an indirectly heated vacuum tube diode.
From top to bottom, the element names are: *plate*, *cathode*, and *heater*.

In operation, the cathode is heated to red heat (800–1000 °C). A directly heated cathode is made of tungsten wire and is heated by current passed through it from an external voltage source. An indirectly heated cathode is heated by infrared radiation from a nearby heater that is formed of Nichrome wire and supplied with current provided by an external voltage source.

The operating temperature of the cathode causes it to release electrons into the vacuum, a process called thermionic emission. The cathode is coated with oxides of alkaline earth metals, such as barium and strontium oxides. These have a low work function, meaning that they more readily emit electrons than would the uncoated cathode.

The plate, not being heated, does not emit electrons; but is able to absorb them.

The alternating voltage to be rectified is applied between the cathode and the plate. When the plate voltage is positive with respect to the cathode, the plate electrostatically attracts the electrons from the cathode, so a current of electrons flows through the tube from cathode to plate. When the plate voltage is negative with respect to the cathode, no electrons are emitted by the plate, so no current can pass from the plate to the cathode.

Semiconductor Diodes

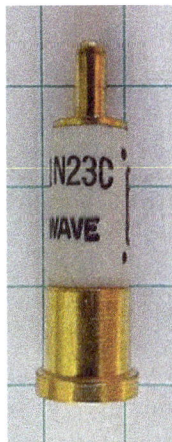

Point contact diode (*crystal rectifier* or *crystal diode*), type 1N23C. Grid one quarter inch.

Point-contact Diodes

Point-contact diodes were developed starting in the 1930s, out of the early crystal detector technology, and are now generally used in the 3 to 30 gigahertz range. Point-contact diodes use a small diameter metal wire in contact with a semiconductor crystal, and are of either *non-welded* contact type or *welded contact* type. Non-welded contact construction utilizes the Schottky barrier principle. The metal side is the pointed end of a small diameter wire that is in contact with the semiconductor crystal. In the welded contact type, a small P region is formed in the otherwise N type crystal around the metal point during manufacture by momentarily passing a relatively large current through the device. Point contact diodes generally exhibit lower capacitance, higher forward resistance and greater reverse leakage than junction diodes.

Junction Diodes

p–n Junction Diode

A p–n junction diode is made of a crystal of semiconductor, usually silicon, but germanium and gallium arsenide are also used. Impurities are added to it to create a region on one side that contains negative charge carriers (electrons), called an n-type semiconductor, and a region on the other side that contains positive charge carriers (holes), called a p-type semiconductor. When the n-type and p-type materials are attached together, a momentary flow of electrons occur from the n to the p side resulting in a third region between the two where no charge carriers are present. This region is called the depletion region because there are no charge carriers (neither electrons nor holes) in it. The diode's terminals are attached to the n-type and p-type regions. The boundary between these two regions, called a p–n junction, is where the action of the diode takes place. When a sufficiently higher electrical potential is applied to the P side (the anode) than to the N side (the cathode), it allows electrons to flow through the depletion region from the N-type side to the P-type side. The junction does not allow the flow of electrons in the opposite direction when the potential is applied in reverse, creating, in a sense, an electrical check valve.

Schottky Diode

Another type of junction diode, the Schottky diode, is formed from a metal–semiconductor junction rather than a p–n junction, which reduces capacitance and increases switching speed.

Current–voltage Characteristic

A semiconductor diode's behavior in a circuit is given by its current–voltage characteristic, or I–V graph. The shape of the curve is determined by the transport of charge carriers through the so-called depletion layer or depletion region that exists at the p–n junction between differing semiconductors. When a p–n junction is first created, conduction-band (mobile) electrons from the N-doped region diffuse into the P-doped region where there is a large population of holes (vacant places for electrons) with which the electrons "recombine". When a mobile electron recombines with a hole, both hole and electron vanish, leaving behind an immobile positively charged donor (dopant) on the N side and negatively charged acceptor (dopant) on the P side. The region around the p–n junction becomes depleted of charge carriers and thus behaves as an insulator.

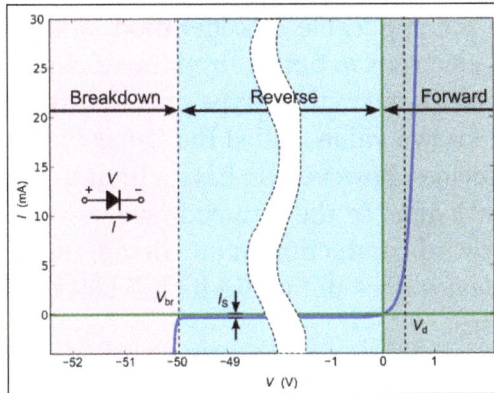

I–V (current vs. voltage) characteristics of a p–n junction diode.

However, the width of the depletion region (called the depletion width) cannot grow without limit. For each electron–hole pair recombination made, a positively charged dopant ion is left behind in the N-doped region, and a negatively charged dopant ion is created in the P-doped region. As recombination proceeds and more ions are created, an increasing electric field develops through the depletion zone that acts to slow and then finally stop recombination. At this point, there is a "built-in" potential across the depletion zone.

Reverse Bias

If an external voltage is placed across the diode with the same polarity as the built-in potential, the depletion zone continues to act as an insulator, preventing any significant electric current flow (unless electron–hole pairs are actively being created in the junction by, for instance, light;). This is called the *reverse bias* phenomenon.

Forward Bias

However, if the polarity of the external voltage opposes the built-in potential, recombination can once again proceed, resulting in a substantial electric current through the p–n junction (i.e. substantial numbers of electrons and holes recombine at the junction). For silicon diodes, the built-in potential is approximately 0.7 V (0.3 V for germanium and 0.2 V for Schottky). Thus, if an external voltage greater than and opposite to the built-in voltage is applied, a current will flow and the diode is said to be "turned on" as it has been given an external *forward bias*. The diode is commonly said to have a forward "threshold" voltage, above which it conducts and below which conduction stops. However, this is only an approximation as the forward characteristic is smooth.

A diode's I–V characteristic can be approximated by four regions of operation:

1. At very large reverse bias, beyond the peak inverse voltage or PIV, a process called reverse breakdown occurs that causes a large increase in current (i.e., a large number of electrons and holes are created at, and move away from the p–n junction) that usually damages the device permanently. The avalanche diode is deliberately designed for use in that manner. In the Zener diode, the concept of PIV is not applicable. A Zener diode contains a heavily doped p–n junction allowing electrons to tunnel from the valence band of the p-type material to the conduction band of the n-type material, such that the reverse voltage is "clamped" to a known value (called the *Zener voltage*), and avalanche does not occur. Both devices, however, do have a limit to the maximum current and power they can withstand in the clamped reverse-voltage region. Also, following the end of forward conduction in any diode, there is reverse current for a short time. The device does not attain its full blocking capability until the reverse current ceases.

2. For a bias less than the PIV, the reverse current is very small. For a normal P–N rectifier diode, the reverse current through the device in the micro-ampere (μA) range is very low. However, this is temperature dependent, and at sufficiently high temperatures, a substantial amount of reverse current can be observed (mA or more). There is also a tiny surface leakage current caused by electrons simply going around the diode as though it were an imperfect insulator.

3. With a small forward bias, where only a small forward current is conducted, the current–voltage curve is exponential in accordance with the ideal diode equation. There is a definite forward voltage at which the diode starts to conduct significantly. This is called the *knee voltage* or *cut-in voltage* and is equal to the barrier potential of the p-n junction. This is a feature of the exponential curve, and appears sharper on a current scale more compressed.

4. At larger forward currents the current-voltage curve starts to be dominated by the ohmic resistance of the bulk semiconductor. The curve is no longer exponential, it is asymptotic to a straight line whose slope is the bulk resistance. This region is particularly important for power diodes. The diode can be modeled as an ideal diode in series with a fixed resistor.

In a small silicon diode operating at its rated currents, the voltage drop is about 0.6 to 0.7 volts. The value is different for other diode types—Schottky diodes can be rated as low as 0.2 V, germanium diodes 0.25 to 0.3 V, and red or blue light-emitting diodes (LEDs) can have values of 1.4 V and 4.0 V respectively.

At higher currents the forward voltage drop of the diode increases. A drop of 1 V to 1.5 V is typical at full rated current for power diodes.

Shockley Diode Equation

The Shockley ideal diode equation or the diode law (named after the bipolar junction transistor co-inventor William Bradford Shockley) gives the I–V characteristic of an ideal diode in either forward or reverse bias (or no bias). The following equation is called the Shockley ideal diode equation when n, the ideality factor, is set equal to 1:

$$I = I_S \left(e^{\frac{V_D}{nV_T}} - 1 \right)$$

where,

I is the diode current,

I_S is the reverse bias saturation current (or scale current),

V_D is the voltage across the diode,

V_T is the thermal voltage, and

n is the *ideality factor*, also known as the *quality factor* or sometimes *emission coefficient*. The ideality factor n typically varies from 1 to 2 (though can in some cases be higher), depending on the fabrication process and semiconductor material and is set equal to 1 for the case of an "ideal" diode (thus the n is sometimes omitted). The ideality factor was added to account for imperfect junctions as observed in real transistors. The factor mainly accounts for carrier recombination as the charge carriers cross the depletion region.

The thermal voltage V_T is approximately 25.85 mV at 300 K, a temperature close to "room temperature" commonly used in device simulation software. At any temperature it is a known constant defined by:

$$V_T = \frac{kT}{q},$$

where k is the Boltzmann constant, T is the absolute temperature of the p–n junction, and q is the magnitude of charge of an electron (the elementary charge).

The reverse saturation current, I_S, is not constant for a given device, but varies with temperature; usually more significantly than V_T, so that V_D typically decreases as T increases.

The *Shockley ideal diode equation* or the *diode law* is derived with the assumption that the only processes giving rise to the current in the diode are drift (due to electrical field), diffusion, and thermal recombination–generation (R–G) (this equation is derived by setting n = 1 above). It also assumes that the R–G current in the depletion region is insignificant. This means that the *Shockley ideal diode equation* doesn't account for the processes involved in reverse breakdown and photon-assisted R–G. Additionally, it doesn't describe the "leveling off" of the I–V curve at high forward bias due to internal resistance. Introducing the ideality factor, n, accounts for recombination and generation of carriers.

Under *reverse bias* voltages the exponential in the diode equation is negligible, and the current is a constant (negative) reverse current value of $-I_S$. The reverse *breakdown region* is not modeled by the Shockley diode equation.

For even rather small *forward bias* voltages the exponential is very large, since the thermal voltage is very small in comparison. The subtracted '1' in the diode equation is then negligible and the forward diode current can be approximated by,

$$I = I_S e^{\frac{V_D}{nV_T}}$$

Small-signal Behavior

At forward voltages less than the saturation voltage, the voltage versus current characteristic curve of most diodes is not a straight line. The current can be approximated by,

$$I = I_S e^{\frac{V_D}{nV_T}}$$

In detector and mixer applications, the current can be estimated by a Taylor's series. The odd terms can be omitted because they produce frequency components that are outside the pass band of the mixer or detector. Even terms beyond the second derivative usually need not be included because they are small compared to the second order term. The desired current component is approximately proportional to the square of the input voltage, so the response is called *square law* in this region.

Reverse-recovery Effect

Following the end of forward conduction in a p–n type diode, a reverse current can flow for a short time. The device does not attain its blocking capability until the mobile charge in the junction is depleted.

The effect can be significant when switching large currents very quickly. A certain amount of "reverse recovery time" t_r (on the order of tens of nanoseconds to a few microseconds) may be required to remove the reverse recovery charge Q_r from the diode. During this recovery time, the diode can actually conduct in the reverse direction. This might give rise to a large constant current in the reverse direction for a short time while the diode is reverse biased. The magnitude of such a reverse current is determined by the operating circuit (i.e., the series resistance) and the diode is said to be in the storage-phase. In certain real-world cases it is important to consider the losses that are incurred by this non-ideal diode effect. However, when the slew rate of the current is not so severe (e.g. Line frequency) the effect can be safely ignored. For most applications, the effect is also negligible for Schottky diodes.

The reverse current ceases abruptly when the stored charge is depleted; this abrupt stop is exploited in step recovery diodes for generation of extremely short pulses.

Types of Semiconductor Diode

Normal (p–n) diodes, which operate are usually made of doped silicon or germanium. Before the development of silicon power rectifier diodes, cuprous oxide and later selenium was used. Their low efficiency required a much higher forward voltage to be applied (typically 1.4 to 1.7 V per "cell", with multiple cells stacked so as to increase the peak inverse voltage rating for application in high voltage rectifiers), and required a large heat sink (often an extension of the diode's metal substrate), much larger than the later silicon diode of the same current ratings would require. The vast majority of all diodes are the p–n diodes found in CMOS integrated circuits, which include two diodes per pin and many other internal diodes.

Avalanche Diodes

These are diodes that conduct in the reverse direction when the reverse bias voltage exceeds the breakdown voltage. These are electrically very similar to Zener diodes (and are often mistakenly called Zener diodes), but break down by a different mechanism: the *avalanche effect*. This occurs when the reverse electric field applied across the p–n junction causes a wave of ionization, reminiscent of an avalanche, leading to a large current. Avalanche diodes are designed to break down at a well-defined reverse voltage without being destroyed. The difference between the avalanche diode (which has a reverse breakdown above about 6.2 V) and the Zener is that the channel length of the former exceeds the mean free path of the electrons, resulting in many collisions between them on the way through the channel. The only practical difference between the two types is they have temperature coefficients of opposite polarities.

Constant Current Diodes

These are actually JFETs with the gate shorted to the source, and function like a two-terminal current-limiting analog to the voltage-limiting Zener diode. They allow a current through them to rise to a certain value, and then level off at a specific value. Also called CLDs, constant-current diodes, diode-connected transistors, or current-regulating diodes.

Crystal Rectifiers or Crystal Diodes

These are point-contact diodes. The 1N21 series and others are used in mixer and detector applications in radar and microwave receivers. The 1N34A is another example of a crystal diode.

Gunn Diodes

These are similar to tunnel diodes in that they are made of materials such as GaAs or InP that exhibit a region of negative differential resistance. With appropriate biasing, dipole domains form and travel across the diode, allowing high frequency microwave oscillators to be built.

Light-emitting Diodes (LEDs)

In a diode formed from a direct band-gap semiconductor, such as gallium arsenide, charge carriers that cross the junction emit photons when they recombine with the majority carrier on the other side. Depending on the material, wavelengths (or colors) from the infrared to the near ultraviolet may be produced. The first LEDs were red and yellow, and higher-frequency diodes have been developed over time. All LEDs produce incoherent, narrow-spectrum light; "white" LEDs are actually a blue LED with a yellow scintillator coating, or combinations of three LEDs of a different color. LEDs can also be used as low-efficiency photodiodes in signal applications. An LED may be paired with a photodiode or phototransistor in the same package, to form an opto-isolator.

Laser Diodes

When an LED-like structure is contained in a resonant cavity formed by polishing the parallel end faces, a laser can be formed. Laser diodes are commonly used in optical storage devices and for high speed optical communication.

Thermal Diodes

This term is used both for conventional p–n diodes used to monitor temperature because of their varying forward voltage with temperature, and for Peltier heat pumps for thermoelectric heating and cooling. Peltier heat pumps may be made from semiconductor, though they do not have any rectifying junctions, they use the differing behaviour of charge carriers in N and P type semiconductor to move heat.

Photodiodes

All semiconductors are subject to optical charge carrier generation. This is typically an undesired effect, so most semiconductors are packaged in light blocking material. Photodiodes are intended to sense light (photodetector), so they are packaged in materials that allow light to pass, and are usually PIN (the kind of diode most sensitive to light).

A photodiode can be used in solar cells, in photometry, or in optical communications. Multiple photodiodes may be packaged in a single device, either as a linear array or as a two-dimensional array. These arrays should not be confused with charge-coupled devices.

PIN Diodes

A PIN diode has a central un-doped, or *intrinsic*, layer, forming a p-type/intrinsic/n-type structure. They are used as radio frequency switches and attenuators. They are also used as large-volume, ionizing-radiation detectors and as photodetectors. PIN diodes are also used in power electronics, as their central layer can withstand high voltages. Furthermore, the PIN structure can be found in many power semiconductor devices, such as IGBTs, power MOSFETs, and thyristors.

Schottky Diodes

Schottky diodes are constructed from a metal to semiconductor contact. They have a lower forward voltage drop than p–n junction diodes. Their forward voltage drop at forward currents of about 1 mA is in the range 0.15 V to 0.45 V, which makes them useful in voltage clamping applications and prevention of transistor saturation. They can also be used as low loss rectifiers, although their reverse leakage current is in general higher than that of other diodes. Schottky diodes are majority carrier devices and so do not suffer from minority carrier storage problems that slow down many other diodes— so they have a faster reverse recovery than p–n junction diodes. They also tend to have much lower junction capacitance than p–n diodes, which provides for high switching speeds and their use in high-speed circuitry and RF devices such as switched-mode power supply, mixers, and detectors.

Super Barrier Diodes

Super barrier diodes are rectifier diodes that incorporate the low forward voltage drop of the Schottky diode with the surge-handling capability and low reverse leakage current of a normal p–n junction diode.

Gold-doped Diodes

As a dopant, gold (or platinum) acts as recombination centers, which helps a fast recombination of minority carriers. This allows the diode to operate at signal frequencies, at the expense of a higher forward voltage drop. Gold-doped diodes are faster than other p–n diodes (but not as fast as Schottky diodes). They also have less reverse-current leakage than Schottky diodes (but not as good as other p–n diodes). A typical example is the 1N914.

Snap-off or Step Recovery Diodes

The term *step recovery* relates to the form of the reverse recovery characteristic of these devices. After a forward current has been passing in an SRD and the current is interrupted or reversed, the reverse conduction will cease very abruptly (as in a step waveform). SRDs can, therefore, provide very fast voltage transitions by the very sudden disappearance of the charge carriers.

Stabistors or Forward Reference Diodes

The term *stabistor* refers to a special type of diodes featuring extremely stable forward voltage characteristics. These devices are specially designed for low-voltage stabilization applications requiring a guaranteed voltage over a wide current range and highly stable over temperature.

Transient Voltage Suppression Diode (TVS)

These are avalanche diodes designed specifically to protect other semiconductor devices from high-voltage transients. Their p–n junctions have a much larger cross-sectional area than those of a normal diode, allowing them to conduct large currents to ground without sustaining damage.

Tunnel Diodes or Esaki Diodes

These have a region of operation showing negative resistance caused by quantum tunneling, allowing amplification of signals and very simple bistable circuits. Because of the high carrier concentration, tunnel diodes are very fast, may be used at low (mK) temperatures, high magnetic fields, and in high radiation environments. Because of these properties, they are often used in spacecraft.

Varicap or Varactor Diodes

These are used as voltage-controlled capacitors. These are important in PLL (phase-locked loop) and FLL (frequency-locked loop) circuits, allowing tuning circuits, such as those in television receivers, to lock quickly on to the frequency. They also enabled tunable oscillators in early discrete tuning of radios, where a cheap and stable, but fixed-frequency, crystal oscillator provided the reference frequency for a voltage-controlled oscillator.

Zener Diodes

These can be made to conduct in reverse bias (backward), and are correctly termed reverse breakdown diodes. This effect, called Zener breakdown, occurs at a precisely defined voltage, allowing the diode to be used as a precision voltage reference. The term Zener diode is colloquially applied to several types of breakdown diodes, but strictly speaking Zener diodes have a breakdown voltage of below 5 volts, whilst avalanche diodes are used for breakdown voltages above that value. In practical voltage reference circuits, Zener and switching diodes are connected in series and opposite directions to balance the temperature coefficient response of the diodes to near-zero. Some devices labeled as high-voltage Zener diodes are actually avalanche diodes. Two (equivalent) Zeners in series and in reverse order, in the same package, constitute a transient absorber (or Transorb, a registered trademark).

Other uses for semiconductor diodes include the sensing of temperature, and computing analog logarithms.

Graphic Symbols

The symbol used to represent a particular type of diode in a circuit diagram conveys the general electrical function to the reader. There are alternative symbols for some types of diodes, though the differences are minor. The triangle in the symbols points to the forward direction, i.e. in the direction of conventional current flow.

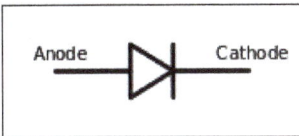

Diode.

Light-emitting diode (LED).

Photodiode.

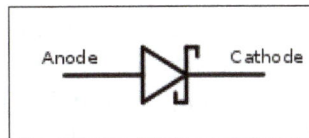

Schottky diode.

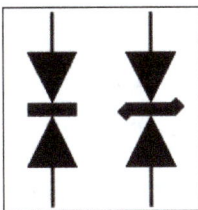

Transient-voltage-suppression diode (TVS).

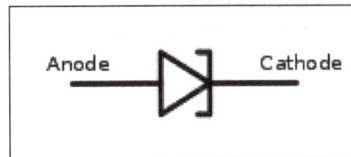

Tunnel diode.

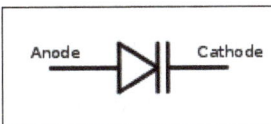

Varicap.

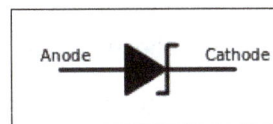

Zener diode.

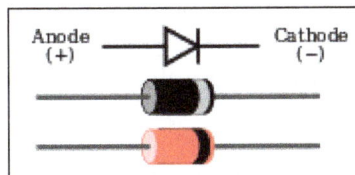

Typical diode packages in same alignment as diode symbol. Thin bar depicts the cathode.

Numbering and Coding Schemes

There are a number of common, standard and manufacturer-driven numbering and coding schemes for diodes; the two most common being the EIA/JEDEC standard and the European Pro Electron standard.

EIA/JEDEC

The standardized 1N-series numbering *EIA370* system was introduced in the US by EIA/JEDEC (Joint Electron Device Engineering Council) about 1960. Most diodes have a 1-prefix designation (e.g., 1N4003). Among the most popular in this series were: 1N34A/1N270 (germanium signal), 1N914/1N4148 (silicon signal), 1N400x (silicon 1A power rectifier), and 1N580x (silicon 3A power rectifier).

JIS

The JIS semiconductor designation system has all semiconductor diode designations starting with "1S".

Pro Electron

The European Pro Electron coding system for active components was introduced in 1966 and comprises two letters followed by the part code. The first letter represents the semiconductor material used for the component (A = germanium and B = silicon) and the second letter represents the general function of the part (for diodes, A = low-power/ signal, B = variable capacitance, X = multiplier, Y = rectifier and Z = voltage reference); for example:

- AA-series germanium low-power/signal diodes (e.g., AA119).
- BA-series silicon low-power/signal diodes (e.g., BAT18 silicon RF switching diode).
- BY-series silicon rectifier diodes (e.g., BY127 1250V, 1A rectifier diode).
- BZ-series silicon Zener diodes (e.g., BZY88C4V7 4.7V Zener diode).

Other common numbering / coding systems (generally manufacturer-driven) include:

- GD-series germanium diodes (e.g., GD9) – this is a very old coding system.
- OA-series germanium diodes (e.g., OA47) – a coding sequence developed by Mullard, a UK company.

As well as these common codes, many manufacturers or organisations have their own systems too – for example:

- HP diode 1901-0044 = JEDEC 1N4148.
- UK military diode CV448 = Mullard type OA81 = GEC type GEX23.

Applications

Radio Demodulation

The first use for the diode was the demodulation of amplitude modulated (AM) radio broadcasts. In summary, an AM signal consists of alternating positive and negative peaks of a radio carrier wave, whose amplitude or envelope is proportional to the original audio signal. The diode rectifies the AM radio frequency signal, leaving only the positive peaks of the carrier wave. The audio is then extracted from the rectified carrier wave using a simple filter and fed into an audio amplifier or transducer, which generates sound waves.

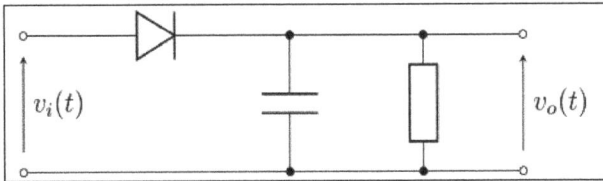

A simple envelope demodulator circuit.

In microwave and millimeter wave technology, beginning in the 1930s, researchers improved and miniaturized the crystal detector. Point contact diodes (*crystal diodes*) and Schottky diodes are used in radar, microwave and millimeter wave detectors.

Power Conversion

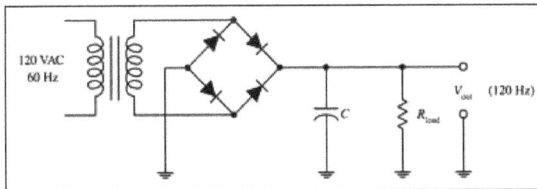

Schematic of basic ac-to-dc power supply.

Rectifiers are constructed from diodes, where they are used to convert alternating current (ac) electricity into direct current (dc). Automotive alternators are a common example, where the diode, which rectifies the AC into dc, provides better performance than the commutator or earlier, dynamo. Similarly, diodes are also used in *Cockcroft–Walton voltage multipliers* to convert ac into higher ac voltages.

Over-voltage Protection

Diodes are frequently used to conduct damaging high voltages away from sensitive electronic devices. They are usually reverse-biased (non-conducting) under normal circumstances. When the voltage rises above the normal range, the diodes become forward-biased (conducting). For example, diodes are used in (stepper motor and H-bridge) motor controller and relay circuits to de-energize coils rapidly without the damaging voltage spikes that would otherwise occur. (A diode used in such an application is called a flyback diode). Many integrated circuits also incorporate diodes on the connection pins to prevent external voltages from damaging their sensitive transistors. Specialized diodes are used to protect from over-voltages at higher power.

Logic Gates

Diodes can be combined with other components to construct AND and OR logic gates. This is referred to as diode logic.

Ionizing Radiation Detectors

In addition to light, semiconductor diodes are sensitive to more energetic radiation. In electronics, cosmic rays and other sources of ionizing radiation cause noise pulses and single and multiple bit errors. This effect is sometimes exploited by particle detectors to detect radiation. A single particle of radiation, with thousands or millions of electron volts of energy, generates many charge carrier pairs, as its energy is deposited in the semiconductor material. If the depletion layer is large enough to catch the whole shower or to stop a heavy particle, a fairly accurate measurement of the particle's energy can be made, simply by measuring the charge conducted and without the complexity of a magnetic spectrometer, etc. These semiconductor radiation detectors need efficient and uniform charge collection and low leakage current. They are often cooled by liquid nitrogen. For longer-range (about a centimetre) particles, they need a very large depletion depth and large area. For short-range particles, they need any contact or un-depleted semiconductor on at least one surface to be very thin. The back-bias voltages are near breakdown (around a thousand volts per centimetre). Germanium and silicon are common materials. Some of these detectors sense position as well as energy. They have a finite life, especially when detecting heavy particles, because of radiation damage. Silicon and germanium are quite different in their ability to convert gamma rays to electron showers.

Semiconductor detectors for high-energy particles are used in large numbers. Because of energy loss fluctuations, accurate measurement of the energy deposited is of less use.

Temperature Measurements

A diode can be used as a temperature measuring device, since the forward voltage drop across the diode depends on temperature, as in a silicon bandgap temperature sensor. From the Shockley ideal diode equation given above, it might *appear* that the voltage has a *positive* temperature coefficient (at a constant current), but usually the variation of the reverse saturation current term is more significant than the variation in the thermal voltage term. Most diodes therefore have a *negative* temperature coefficient, typically −2 mV/°C for silicon diodes. The temperature coefficient is approximately constant for temperatures above about 20 kelvin. Some graphs are given for 1N400x series, and CY7 cryogenic temperature sensor.

Current Steering

Diodes will prevent currents in unintended directions. To supply power to an electrical circuit during a power failure, the circuit can draw current from a battery. An uninterruptible power supply may use diodes in this way to ensure that current is only drawn from the battery when necessary. Likewise, small boats typically have two circuits each with their own battery/batteries: one used for engine starting; one used for domestics. Normally, both are charged from a single alternator, and a heavy-duty split-charge diode is used to prevent the higher-charge battery (typically the engine battery) from discharging through the lower-charge battery when the alternator is not running.

Diodes are also used in electronic musical keyboards. To reduce the amount of wiring needed in electronic musical keyboards, these instruments often use keyboard matrix circuits. The keyboard controller scans the rows and columns to determine which note the player has pressed. The problem with matrix circuits is that, when several notes are pressed at once, the current can flow backwards through the circuit and trigger "phantom keys" that cause "ghost" notes to play. To avoid triggering unwanted notes, most keyboard matrix circuits have diodes soldered with the switch under each key of the musical keyboard. The same principle is also used for the switch matrix in solid-state pinball machines.

Waveform Clipper

Diodes can be used to limit the positive or negative excursion of a signal to a prescribed voltage.

Clamper

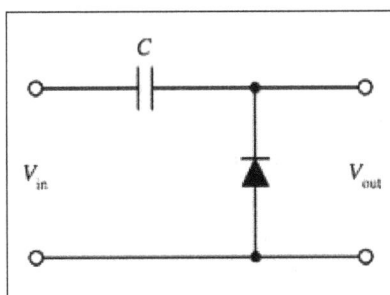

This simple diode clamp will clamp the negative peaks
of the incoming waveform to the common rail voltage.

A diode clamp circuit can take a periodic alternating current signal that oscillates between positive and negative values, and vertically displace it such that either the positive, or the negative peaks occur at a prescribed level. The clamper does not restrict the peak-to-peak excursion of the signal, it moves the whole signal up or down so as to place the peaks at the reference level.

Abbreviations

Diodes are usually referred to as D for diode on PCBs. Sometimes the abbreviation CR for *crystal rectifier* is used.

ZENER DIODE

A Zener diode is a type of diode that allows current to flow not only from its anode to its cathode, but also in the reverse direction, when the Zener voltage is reached.

Zener diodes have a highly doped p–n junction. Normal diodes break down with a reverse voltage, but the voltage and sharpness of the knee are not as well defined as for a Zener diode. Normal diodes are not designed to operate in the breakdown region, whereas Zener diodes operate reliably in this region.

Zener reverse breakdown is due to electron quantum tunnelling caused by a high-strength electric field. However, many diodes described as "Zener" diodes rely instead on avalanche breakdown. Both breakdown types are used in Zener diodes with the Zener effect predominating at lower voltages and avalanche breakdown at higher voltages.

Zener diodes are widely used in electronic equipment of all kinds and are one of the basic building blocks of electronic circuits. They are used to generate low-power stabilized supply rails from a higher voltage and to provide reference voltages for circuits, especially stabilized power supplies. They are also used to protect circuits from overvoltage, especially electrostatic discharge (ESD).

Operation

A conventional solid-state diode allows significant current if it is reverse-biased above its reverse breakdown voltage. When the reverse bias breakdown voltage is exceeded, a conventional diode is subject to high current due to avalanche breakdown. Unless this current is limited by circuitry, the diode may be permanently damaged due to overheating. A Zener diode exhibits almost the same properties, except the device is specially designed so as to have a reduced breakdown voltage, the so-called Zener voltage. By contrast with the conventional device, a reverse-biased Zener diode exhibits a controlled breakdown and allows the current to keep the voltage across the Zener diode close to the Zener breakdown voltage. For example, a diode with a Zener breakdown voltage of 3.2 V exhibits a voltage drop of very nearly 3.2 V across a wide range of reverse currents. The Zener diode is therefore ideal for applications such as the generation of a reference voltage (e.g. for an amplifier stage), or as a voltage stabilizer for low-current applications.

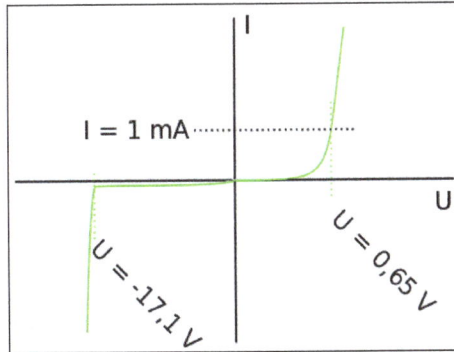

Current-voltage characteristic of a Zener diode with a breakdown voltage of 17 V. Notice the change of voltage scale between the forward biased (positive) direction and the reverse biased (negative) direction.

Another mechanism that produces a similar effect is the avalanche effect as in the avalanche diode. The two types of diode are in fact constructed the same way and both effects are present in diodes of this type. In silicon diodes up to about 5.6 volts, the Zener effect is the predominant effect and shows a marked negative temperature coefficient. Above 5.6 volts, the avalanche effect becomes predominant and exhibits a positive temperature coefficient.

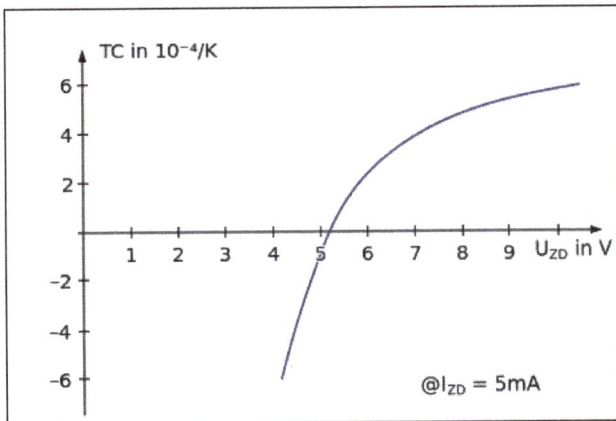

Temperature coefficient of Zener voltage against nominal Zener voltage.

In a 5.6 V diode, the two effects occur together, and their temperature coefficients nearly cancel each other out, thus the 5.6 V diode is useful in temperature-critical applications. An alternative, which is used for voltage references that need to be highly stable over long periods of time, is to use a Zener diode with a temperature coefficient (TC) of +2 mV/°C (breakdown voltage 6.2–6.3 V) connected in series with a forward-biased silicon diode (or a transistor B-E junction) manufactured on the same chip. The forward-biased diode has a temperature coefficient of –2 mV/°C, causing the TCs to cancel out.

Modern manufacturing techniques have produced devices with voltages lower than 5.6 V with negligible temperature coefficients, but as higher-voltage devices are encountered, the temperature coefficient rises dramatically. A 75 V diode has 10 times the coefficient of a 12 V diode.

Zener and avalanche diodes, regardless of breakdown voltage, are usually marketed under the umbrella term of "Zener diode".

Under 5.6 V, where the Zener effect dominates, the IV curve near breakdown is much more rounded, which calls for more care in targeting its biasing conditions. The IV curve for Zeners above 5.6 V (being dominated by Avalanche), is much sharper at breakdown.

Construction

The Zener diode's operation depends on the heavy doping of its p-n junction. The depletion region formed in the diode is very thin (<1 μm) and the electric field is consequently very high (about 500 kV/m) even for a small reverse bias voltage of about 5 V, allowing electrons to tunnel from the valence band of the p-type material to the conduction band of the n-type material.

At the atomic scale, this tunneling corresponds to the transport of valence band electrons into the empty conduction band states; as a result of the reduced barrier between these bands and high electric fields that are induced due to the high levels of doping on both sides. The breakdown voltage can be controlled quite accurately in the doping process. While tolerances within 0.07% are available, the most widely used tolerances are 5% and 10%. Breakdown voltage for commonly available Zener diodes can vary widely from 1.2 V to 200 V.

For diodes that are lightly doped the breakdown is dominated by the avalanche effect rather than the Zener effect. Consequently, the breakdown voltage is higher (over 5.6 V) for these devices.

Surface Zeners

The emitter-base junction of a bipolar NPN transistor behaves as a Zener diode, with breakdown voltage at about 6.8 V for common bipolar processes and about 10 V for lightly doped base regions in BiCMOS processes. Older processes with poor control of doping characteristics had the variation of Zener voltage up to ±1 V, newer processes using ion implantation can achieve no more than ±0.25 V. The NPN transistor structure can be employed as a *surface Zener diode*, with collector and emitter connected together as its cathode and base region as anode. In this approach the base doping profile usually narrows towards the surface, creating a region with intensified electric field where the avalanche breakdown occurs. The hot carriers produced by acceleration in the intense field sometime shoot into the oxide layer above the junction and become trapped there. The accumulation of trapped charges can then cause 'Zener walkout', a corresponding change of the Zener voltage of the junction. The same effect can be achieved by radiation damage.

The emitter-base Zener diodes can handle only smaller currents as the energy is dissipated in the base depletion region which is very small. Higher amount of dissipated energy (higher current for longer time, or a short very high current spike) causes thermal damage to the junction and/or its contacts. Partial damage of the junction can shift its Zener voltage. Total destruction of the Zener junction by overheating it and causing migration of metallization across the junction ("spiking") can be used intentionally as a 'Zener zap' antifuse.

Subsurface Zeners

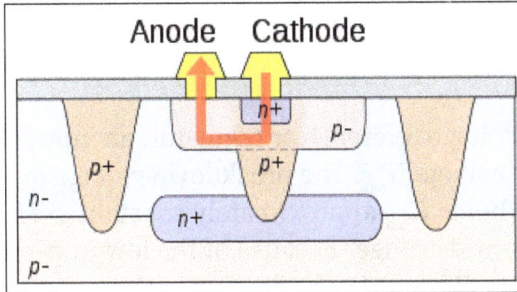

Buried zener structure.

A subsurface Zener diode, also called 'buried Zener', is a device similar to the Surface Zener, but with the avalanche region located deeper in the structure, typically several micrometers below the oxide. The hot carriers then lose energy by collisions with the semiconductor lattice before reaching the oxide layer and cannot be trapped there. The Zener walkout phenomenon therefore does not occur here, and the buried Zeners have voltage constant over their entire lifetime. Most buried Zeners have breakdown voltage of 5–7 volts. Several different junction structures are used.

Uses

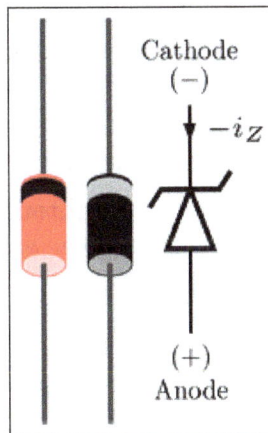

Zener diode shown with typical packages. *Reverse* current $-i_z$ is shown.

Zener diodes are widely used as voltage references and as shunt regulators to regulate the voltage across small circuits. When connected in parallel with a variable voltage source so that it is reverse biased, a Zener diode conducts when the voltage reaches the diode's reverse breakdown voltage. From that point on, the low impedance of the diode keeps the voltage across the diode at that value.

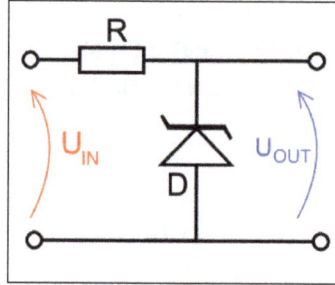

In this circuit, a typical voltage reference or regulator, an input voltage, U_{in}, is regulated down to a stable output voltage U_{out}. The breakdown voltage of diode D is stable over a wide current range and holds U_{out} approximately constant even though the input voltage may fluctuate over a wide range. Because of the low impedance of the diode when operated like this, resistor R is used to limit current through the circuit.

In the case of this simple reference, the current flowing in the diode is determined using Ohm's law and the known voltage drop across the resistor R:

$$I_{diode} = \frac{U_{in} - U_{out}}{R}$$

The value of R must satisfy two conditions:

1. R must be small enough that the current through D keeps D in reverse breakdown. The value of this current is given in the data sheet for D. For example, the common BZX79C5V6 device, a 5.6 V 0.5 W Zener diode, has a recommended reverse current of 5 mA. If insufficient current exists through D, then U_{out} is unregulated and less than the nominal breakdown voltage (this differs from voltage-regulator tubes where the output voltage is higher than nominal and could rise as high as U_{in}). When calculating R, allowance must be made for any current through the external load, not shown in this diagram, connected across U_{out}.

2. R must be large enough that the current through D does not destroy the device. If the current through D is I_D, its breakdown voltage V_B and its maximum power dissipation P_{max} correlate as such: $I_D V_B < P_{max}$.

A load may be placed across the diode in this reference circuit, and as long as the Zener stays in reverse breakdown, the diode provides a stable voltage source to the load. Zener diodes in this configuration are often used as stable references for more advanced voltage regulator circuits.

Shunt regulators are simple, but the requirements that the ballast resistor be small enough to avoid excessive voltage drop during worst-case operation (low input voltage concurrent with high load current) tends to leave a lot of current flowing in the diode much of the time, making for a fairly wasteful regulator with high quiescent power dissipation, only suitable for smaller loads.

These devices are also encountered, typically in series with a base-emitter junction, in transistor stages where selective choice of a device centered on the avalanche or Zener point can be used to introduce compensating temperature co-efficient balancing of the transistor p–n junction. An example of this kind of use would be a DC error amplifier used in a regulated power supply circuit feedback loop system.

Zener diodes are also used in surge protectors to limit transient voltage spikes.

Another application of the Zener diode is the use of noise caused by its avalanche break-down in a random number generator.

Waveform Clipper

Examples of a waveform clipper:

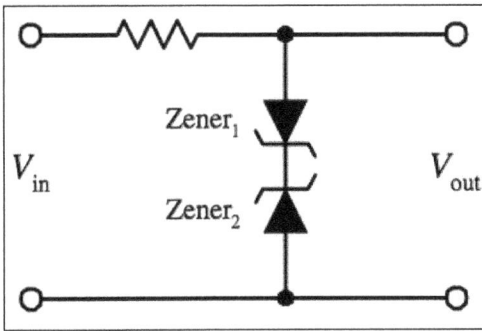

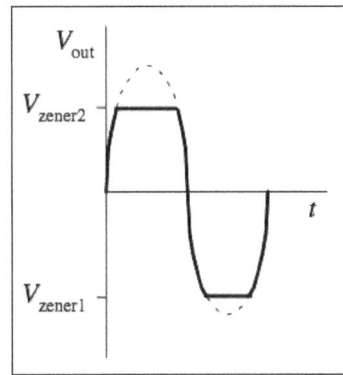

Two Zener diodes facing each other in series clip both halves of an input signal. Waveform clippers can be used not only to reshape a signal, but also to prevent voltage spikes from affecting circuits that are connected to the power supply.

Voltage Shifter

Examples of a voltage shifter:

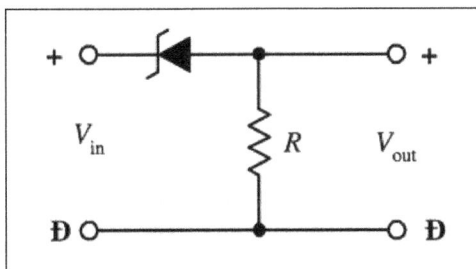

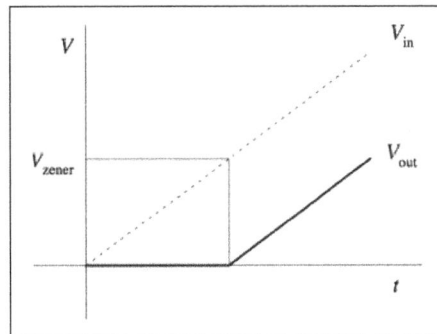

A Zener diode can be applied to a circuit with a resistor to act as a voltage shifter. This circuit lowers the output voltage by a quantity that is equal to the Zener diode's break-down voltage.

Voltage Regulator

Examples of a voltage regulator:

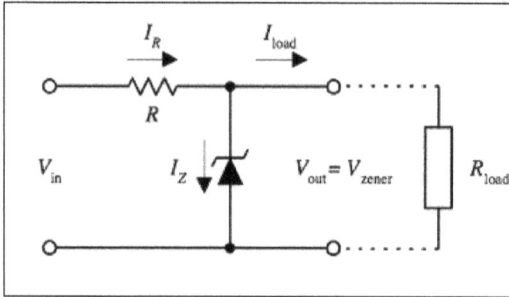

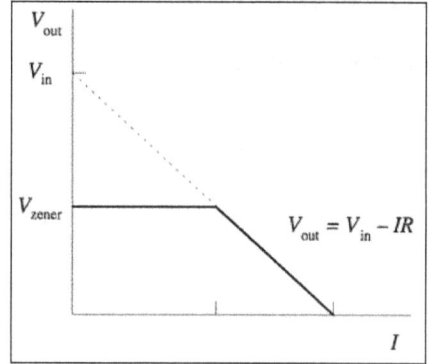

A Zener diode can be applied in a voltage regulator circuit to regulate the voltage applied to a load, such as in a linear regulator.

LIGHT EMITTING DIODE

Light-emitting diode (LED) is a semiconductor light source that emits light when current flows through it. Electrons in the semiconductor recombine with electron holes, releasing energy in the form of photons. The color of the light (corresponding to the energy of the photons) is determined by the energy required for electrons to cross the band gap of the semiconductor. White light is obtained by using multiple semiconductors or a layer of light-emitting phosphor on the semiconductor device.

Appearing as practical electronic components in 1962, the earliest LEDs emitted low-intensity infrared light. Infrared LEDs are used in remote-control circuits, such as those used with a wide variety of consumer electronics. The first visible-light LEDs were of low intensity and limited to red. Modern LEDs are available across the visible, ultraviolet, and infrared wavelengths, with high light output.

Early LEDs were often used as indicator lamps, replacing small incandescent bulbs, and in seven-segment displays. Recent developments have produced high-output white light LEDs suitable for room and outdoor area lighting. LEDs have led to new displays and sensors, while their high switching rates are useful in advanced communications technology.

LEDs have many advantages over incandescent light sources, including lower energy consumption, longer lifetime, improved physical robustness, smaller size, and faster switching. LEDs are used in applications as diverse as aviation lighting, automotive headlamps, advertising, general lighting, traffic signals, camera flashes, lighted wallpaper, plant growing light and medical devices.

Unlike a laser, the color of light emitted from an LED is neither coherent nor monochromatic, but the spectrum is narrow with respect to human vision, and functionally monochromatic.

Physics of Light Production and Emission

In a light emitting diode, the recombination of electrons and electron holes in a semiconductor produces light (or infrared radiation), a process called "electroluminescence". The wavelength of the light depends on the energy band gap of the semiconductors used. Since these materials have a high index of refraction, design features of the devices such as special optical coatings and die shape are required to efficiently emit light.

Colors

By selection of different semiconductor materials, single-color LEDs can be made that emit light in a narrow band of wavelengths from near-infrared through the visible spectrum and into the ultraviolet range. As the wavelengths become shorter, because of the larger band gap of these semiconductors, the operating voltage of the LED increases.

Blue and Ultraviolet

Blue LEDs have an active region consisting of one or more InGaN quantum wells sandwiched between thicker layers of GaN, called cladding layers. By varying the relative In/Ga fraction in the InGaN quantum wells, the light emission can in theory be varied from violet to amber.

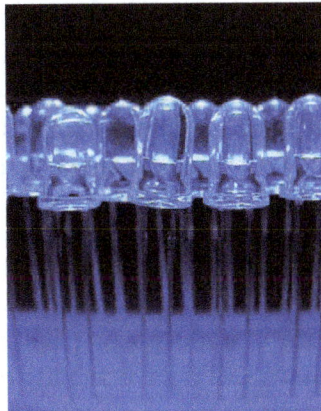

Blue LEDs.

Aluminium gallium nitride (AlGaN) of varying Al/Ga fraction can be used to manufacture the cladding and quantum well layers for ultraviolet LEDs, but these devices have not yet reached the level of efficiency and technological maturity of InGaN/GaN blue/ green devices. If un-alloyed GaN is used in this case to form the active quantum well layers, the device emits near-ultraviolet light with a peak wavelength centred around 365 nm. Green LEDs manufactured from the InGaN/GaN system are far more efficient and brighter than green LEDs produced with non-nitride material systems, but practical devices still exhibit efficiency too low for high-brightness applications.

With AlGaN and AlGaInN, even shorter wavelengths are achievable. Near-UV emitters at wavelengths around 360–395 nm are already cheap and often encountered, for example, as black light lamp replacements for inspection of anti-counterfeiting UV watermarks in documents and bank notes, and for UV curing. While substantially more expensive, shorter-wavelength diodes are commercially available for wavelengths down to 240 nm. As the photosensitivity of microorganisms approximately matches the absorption spectrum of DNA, with a peak at about 260 nm, UV LED emitting at 250–270 nm are expected in prospective disinfection and sterilization devices. commercially available UVA LEDs (365 nm) are already effective disinfection and sterilization devices. UV-C wavelengths were obtained in laboratories using aluminium nitride (210 nm), boron nitride (215 nm) and diamond (235 nm).

White

There are two primary ways of producing white light-emitting diodes. One is to use individual LEDs that emit three primary colors—red, green and blue—and then mix all the colors to form white light. The other is to use a phosphor material to convert monochromatic light from a blue or UV LED to broad-spectrum white light, similar to a fluorescent lamp. The yellow phosphor is cerium-doped YAG crystals suspended in the package or coated on the LED. This YAG phosphor causes white LEDs to look yellow when off.

The 'whiteness' of the light produced is engineered to suit the human eye. Because of metamerism, it is possible to have quite different spectra that appear white. However, the appearance of objects illuminated by that light may vary as the spectrum varies. This is the issue of color rendition, quite separate from color temperature. An orange or cyan object could appear with the wrong color and much darker as the LED or phosphor does not emit the wavelength it reflects. The best color rendition LEDs use a mix of phosphors, resulting in less efficiency but better color rendering.

RGB Systems

Mixing red, green, and blue sources to produce white light needs electronic circuits to control the blending of the colors. Since LEDs have slightly different emission patterns, the color balance may change depending on the angle of view, even if the RGB sources are in a single package, so RGB diodes are seldom used to produce white lighting. Nonetheless, this method has many applications because of the flexibility of mixing different colors, and in principle, this mechanism also has higher quantum efficiency in producing white light.

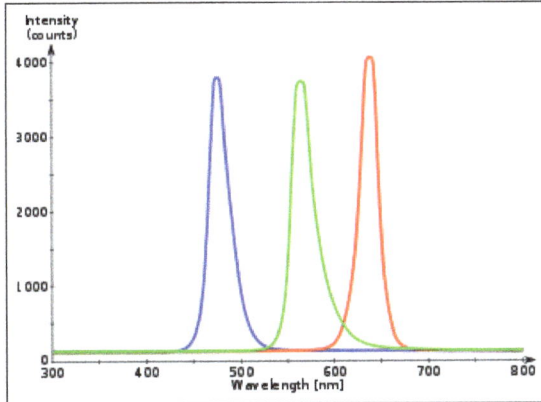

Combined spectral curves for blue, yellow-green, and high-brightness red solid-state semiconductor LEDs. FWHM spectral bandwidth is approximately 24–27 nm for all three colors.

There are several types of multicolor white LEDs: di-, tri-, and tetrachromatic white LEDs. Several key factors that play among these different methods include color stability, color rendering capability, and luminous efficacy. Often, higher efficiency means lower color rendering, presenting a trade-off between the luminous efficacy and color rendering. For example, the dichromatic white LEDs have the best luminous efficacy (120 lm/W), but the lowest color rendering capability. However, although tetrachromatic white LEDs have excellent color rendering capability, they often have poor luminous efficacy. Trichromatic white LEDs are in between, having both good luminous efficacy (>70 lm/W) and fair color rendering capability.

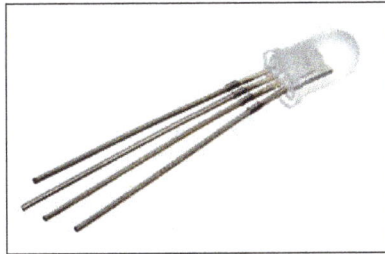

RGB LED.

One of the challenges is the development of more efficient green LEDs. The theoretical maximum for green LEDs is 683 lumens per watt but as of 2010 few green LEDs exceed even 100 lumens per watt. The blue and red LEDs approach their theoretical limits.

Multicolor LEDs also offer a new means to form light of different colors. Most perceivable colors can be formed by mixing different amounts of three primary colors. This allows precise dynamic color control. However, this type of LED's emission power decays exponentially with rising temperature, resulting in a substantial change in color stability. Such problems inhibit industrial use. Multicolor LEDs without phosphors cannot provide good color rendering because each LED is a narrowband source. LEDs without phosphor, while a poorer solution for general lighting, are the best solution for displays, either backlight of LCD, or direct LED based pixels.

Dimming a multicolor LED source to match the characteristics of incandescent lamps is difficult because manufacturing variations, age, and temperature change the actual color value output. To emulate the appearance of dimming incandescent lamps may require a feedback system with color sensor to actively monitor and control the color.

Phosphor-based LEDs

This method involves coating LEDs of one color (mostly blue LEDs made of InGaN) with phosphors of different colors to form white light; the resultant LEDs are called phosphor-based or phosphor-converted white LEDs (pcLEDs). A fraction of the blue light undergoes the Stokes shift, which transforms it from shorter wavelengths to longer. Depending on the original LED's color, various color phosphors are used. Using several phosphor layers of distinct colors broadens the emitted spectrum, effectively raising the color rendering index (CRI).

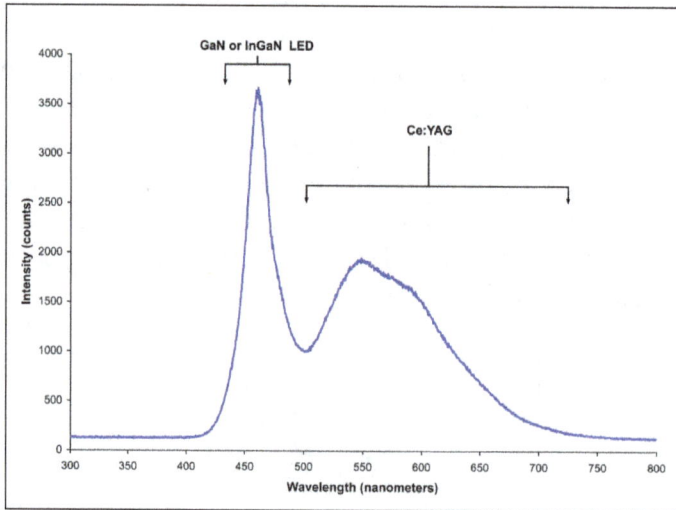

Spectrum of a white LED showing blue light directly emitted by the GaN-based LED (peak at about 465 nm) and the more broadband Stokes-shifted light emitted by the Ce^{3+}:YAG phosphor, which emits at roughly 500–700 nm.

Phosphor-based LEDs have efficiency losses due to heat loss from the Stokes shift and also other phosphor-related issues. Their luminous efficacies compared to normal LEDs depend on the spectral distribution of the resultant light output and the original wavelength of the LED itself. For example, the luminous efficacy of a typical YAG yellow phosphor based white LED ranges from 3 to 5 times the luminous efficacy of the original blue LED because of the human eye's greater sensitivity to yellow than to blue (as modeled in the luminosity function). Due to the simplicity of manufacturing, the phosphor method is still the most popular method for making high-intensity white LEDs. The design and production of a light source or light fixture using a monochrome emitter with phosphor conversion is simpler and cheaper than a complex RGB system, and the majority of high-intensity white LEDs presently on the market are manufactured using phosphor light conversion.

Among the challenges being faced to improve the efficiency of LED-based white light sources is the development of more efficient phosphors. As of 2010, the most efficient yellow phosphor is still the YAG phosphor, with less than 10% Stokes shift loss. Losses attributable to internal optical losses due to re-absorption in the LED chip and in the LED packaging itself account typically for another 10% to 30% of efficiency loss. Currently, in the area of phosphor LED development, much effort is being spent on optimizing these devices to higher light output and higher operation temperatures. For instance, the efficiency can be raised by adapting better package design or by using a more suitable type of phosphor. Conformal coating process is frequently used to address the issue of varying phosphor thickness.

Some phosphor-based white LEDs encapsulate InGaN blue LEDs inside phosphor-coated epoxy. Alternatively, the LED might be paired with a remote phosphor, a preformed polycarbonate piece coated with the phosphor material. Remote phosphors provide more diffuse light, which is desirable for many applications. Remote phosphor designs are also more tolerant of variations in the LED emissions spectrum. A common yellow phosphor material is cerium-doped yttrium aluminium garnet (Ce^{3+}:YAG).

White LEDs can also be made by coating near-ultraviolet (NUV) LEDs with a mixture of high-efficiency europium-based phosphors that emit red and blue, plus copper and aluminium-doped zinc sulfide (ZnS:Cu, Al) that emits green. This is a method analogous to the way fluorescent lamps work. This method is less efficient than blue LEDs with YAG:Ce phosphor, as the Stokes shift is larger, so more energy is converted to heat, but yields light with better spectral characteristics, which render color better. Due to the higher radiative output of the ultraviolet LEDs than of the blue ones, both methods offer comparable brightness. A concern is that UV light may leak from a malfunctioning light source and cause harm to human eyes or skin.

Other White LEDs

Another method used to produce experimental white light LEDs used no phosphors at all and was based on homoepitaxially grown zinc selenide (ZnSe) on a ZnSe substrate that simultaneously emitted blue light from its active region and yellow light from the substrate.

A new style of wafers composed of gallium-nitride-on-silicon (GaN-on-Si) is being used to produce white LEDs using 200-mm silicon wafers. This avoids the typical costly sapphire substrate in relatively small 100- or 150-mm wafer sizes. The sapphire apparatus must be coupled with a mirror-like collector to reflect light that would otherwise be wasted. It is predicted that by 2020, 40% of all GaN LEDs will be made with GaN-on-Si. Manufacturing large sapphire material is difficult, while large silicon material is cheaper and more abundant. LED companies shifting from using sapphire to silicon should be a minimal investment.

Organic Light-emitting Diodes (OLEDs)

In an organic light-emitting diode (OLED), the electroluminescent material composing the emissive layer of the diode is an organic compound. The organic material is electrically conductive due to the delocalization of pi electrons caused by conjugation over all or part of the molecule, and the material therefore functions as an organic semiconductor. The organic materials can be small organic molecules in a crystalline phase, or polymers.

The potential advantages of OLEDs include thin, low-cost displays with a low driving voltage, wide viewing angle, and high contrast and color gamut. Polymer LEDs have the added benefit of printable and flexible displays. OLEDs have been used to make visual displays for portable electronic devices such as cellphones, digital cameras, and MP3 players while possible future uses include lighting and televisions.

Types

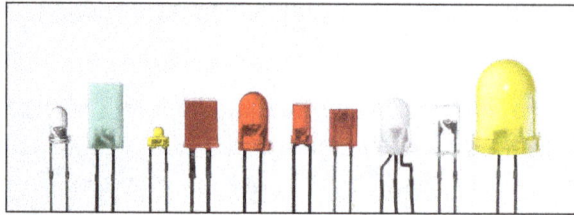

LEDs are produced in a variety of shapes and sizes. The color of the plastic lens is often the same as the actual color of light emitted, but not always. For instance, purple plastic is often used for infrared LEDs, and most blue devices have colorless housings. Modern high-power LEDs such as those used for lighting and backlighting are generally found in surface-mount technology (SMT) packages.

LEDs are made in different packages for different applications. A single or a few LED junctions may be packed in one miniature device for use as an indicator or pilot lamp. An LED array may include controlling circuits within the same package, which may range from a simple resistor, blinking or color changing control, or an addressable controller for RGB devices. Higher-powered white-emitting devices will be mounted on heat sinks and will be used for illumination. Alphanumeric displays in dot matrix or bar formats are widely available. Special packages permit connection of LEDs to optical fibers for high-speed data communication links.

Miniature

These are mostly single-die LEDs used as indicators, and they come in various sizes from 2 mm to 8 mm, through-hole and surface mount packages. Typical current ratings range from around 1 mA to above 20 mA. Multiple LED dies attached to a flexible backing tape form an LED strip light.

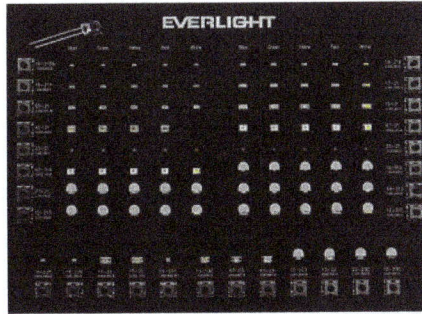

Photo of miniature surface mount LEDs in most common sizes. They can be much smaller than a traditional 5 mm lamp type LED, shown on the upper left corner.

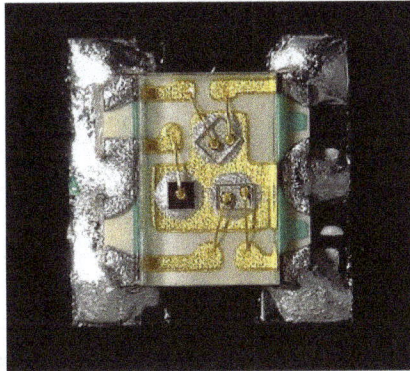

Very small (1.6x1.6x0.35 mm) red, green, and blue surface mount miniature LED package with gold wire bonding details.

Common package shapes include round, with a domed or flat top, rectangular with a flat top (as used in bar-graph displays), and triangular or square with a flat top. The encapsulation may also be clear or tinted to improve contrast and viewing angle. Infrared devices may have a black tint to block visible light while passing infrared radiation.

Ultra-high-output LEDs are designed for viewing in direct sunlight 5 V and 12 V LEDs are ordinary miniature LEDs that have a series resistor for direct connection to a 5 V or 12 V supply.

High-power

High-power light-emitting diodes attached to an LED star base (Luxeon, Lumileds).

High-power LEDs (HP-LEDs) or high-output LEDs (HO-LEDs) can be driven at currents from hundreds of mA to more than an ampere, compared with the tens of mA for other LEDs. Some can emit over a thousand lumens. LED power densities up to 300 W/cm² have been achieved. Since overheating is destructive, the HP-LEDs must be mounted on a heat sink to allow for heat dissipation. If the heat from an HP-LED is not removed, the device fails in seconds. One HP-LED can often replace an incandescent bulb in a flashlight, or be set in an array to form a powerful LED lamp.

Some well-known HP-LEDs in this category are the Nichia 19 series, Lumileds Rebel Led, Osram Opto Semiconductors Golden Dragon, and Cree X-lamp. As of September 2009, some HP-LEDs manufactured by Cree now exceed 105 lm/W.

Examples for Haitz's law—which predicts an exponential rise in light output and efficacy of LEDs over time—are the CREE XP-G series LED, which achieved 105 lm/W in 2009 and the Nichia 19 series with a typical efficacy of 140 lm/W, released in 2010.

AC-driven

LEDs developed by Seoul Semiconductor can operate on AC power without a DC converter. For each half-cycle, part of the LED emits light and part is dark, and this is reversed during the next half-cycle. The efficacy of this type of HP-LED is typically 40 lm/W. A large number of LED elements in series may be able to operate directly from line voltage. In 2009, Seoul Semiconductor released a high DC voltage LED, named as 'Acrich MJT', capable of being driven from AC power with a simple controlling circuit. The low-power dissipation of these LEDs affords them more flexibility than the original AC LED design.

Application-specific Variations

Flashing

Flashing LEDs are used as attention seeking indicators without requiring external electronics. Flashing LEDs resemble standard LEDs but they contain an integrated multivibrator circuit that causes the LED to flash with a typical period of one second. In diffused lens LEDs, this circuit is visible as a small black dot. Most flashing LEDs emit light of one color, but more sophisticated devices can flash between multiple colors and even fade through a color sequence using RGB color mixing.

Bi-color

Bi-color LEDs contain two different LED emitters in one case. There are two types of these. One type consists of two dies connected to the same two leads antiparallel to each other. Current flow in one direction emits one color, and current in the opposite direction emits the other color. The other type consists of two dies with separate leads for both dies and another lead for common anode or cathode so that they can be controlled independently. The most common bi-color combination is red/traditional green, however, other available combinations include amber/traditional green, red/pure green, red/blue, and blue/pure green.

RGB Tri-color

Tri-color LEDs contain three different LED emitters in one case. Each emitter is connected to a separate lead so they can be controlled independently. A four-lead arrangement is typical with one common lead (anode or cathode) and an additional lead for each color. Others, however, have only two leads (positive and negative) and have a built-in electronic controller.

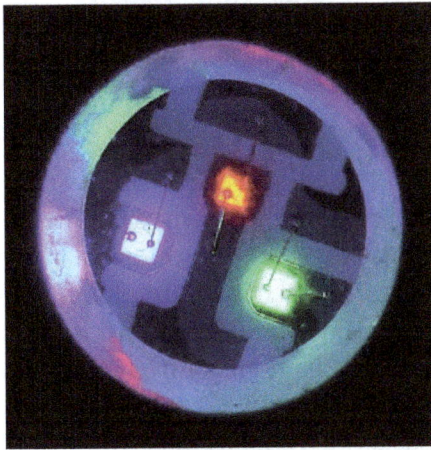

RGB-SMD-LED.

RGB LEDs consist of one red, one green, and one blue LED. By independently adjusting each of the three, RGB LEDs are capable of producing a wide color gamut. Unlike dedicated-color LEDs, however, these do not produce pure wavelengths. Modules may not be optimized for smooth color mixing.

Decorative-multicolor

Decorative-multicolor LEDs incorporate several emitters of different colors supplied by only two lead-out wires. Colors are switched internally by varying the supply voltage.

Alphanumeric

Alphanumeric LEDs are available in seven-segment, starburst, and dot-matrix format. Seven-segment displays handle all numbers and a limited set of letters. Starburst displays can display all letters. Dot-matrix displays typically use 5x7 pixels per character. Seven-segment LED displays were in widespread use in the 1970s and 1980s, but rising use of liquid crystal displays, with their lower power needs and greater display flexibility, has reduced the popularity of numeric and alphanumeric LED displays.

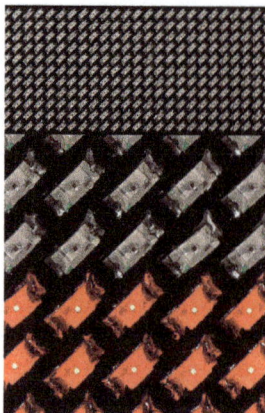

Composite image of an 11 × 44 LED matrix lapel name tag display using 1608/0603-type
SMD LEDs. Top: A little over half of the 21x86 mm display. Center: Close-up
of LEDs in ambient light. Bottom: LEDs in their own red light.

Digital RGB

Digital RGB addressable LEDs contain their own "smart" control electronics. In addition to power and ground, these provide connections for data-in, data-out, and sometimes a clock or strobe signal. These are connected in a daisy chain. Data sent to the first LED of the chain can control the brightness and color of each LED independently of the others. They are used where a combination of maximum control and minimum visible electronics are needed such as strings for Christmas and LED matrices. Some even have refresh rates in the kHz range, allowing for basic video applications. These devices are known by their part number (WS2812 being common) or a brand name such as NeoPixel.

Filament

An LED filament consists of multiple LED chips connected in series on a common longitudinal substrate that forms a thin rod reminiscent of a traditional incandescent filament. These are being used as a low-cost decorative alternative for traditional light bulbs that are being phased out in many countries. The filaments use a rather high voltage, allowing them to work efficiently with mains voltages. Often a simple rectifier and capacitive current limiting are employed to create a low-cost replacement for a traditional light bulb without the complexity of the low voltage, high current converter that single die LEDs need. Usually, they are packaged in bulb similar to the lamps they were designed to replace, and filled with inert gas to remove heat efficiently.

Chip-on-board Arrays

Surface-mounted LEDs are frequently produced in chip on board (COB) arrays, allowing better heat dissipation than with a single LED of comparable luminous output. The LEDs can be arranged around a cylinder, and are called "corn cob lights" because of the rows of yellow LEDs.

Considerations for Use

Power Sources

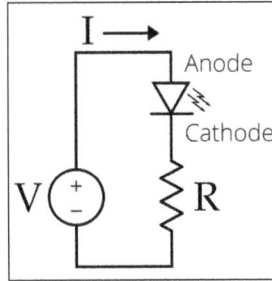

Simple LED circuit with resistor for current limiting.

The current in an LED or other diodes rises exponentially with the applied voltage, so a small change in voltage can cause a large change in current. Current through the LED must be regulated by an external circuit such as a constant current source to prevent damage. Since most common power supplies are (nearly) constant-voltage sources, LED fixtures must include a power converter, or at least a current-limiting resistor. In some applications, the internal resistance of small batteries is sufficient to keep current within the LED rating.

Electrical Polarity

An LED will light only when voltage is applied in the forward direction of the diode. No current flows and no light is emitted if voltage is applied in the reverse direction. If the reverse voltage exceeds the breakdown voltage, a large current flows and the LED will be damaged. If the reverse current is sufficiently limited to avoid damage, the reverse-conducting LED is a useful noise diode.

Safety and Health

Certain blue LEDs and cool-white LEDs can exceed safe limits of the so-called blue-light hazard as defined in eye safety specifications such as "ANSI/IESNA RP-27.1–05: Recommended Practice for Photobiological Safety for Lamp and Lamp Systems". One study showed no evidence of a risk in normal use at domestic illuminance, and that caution is only needed for particular occupational situations or for specific populations.

While LEDs have the advantage over fluorescent lamps, in that they do not contain mercury, they may contain other hazardous metals such as lead and arsenic.

In 2016 the American Medical Association (AMA) issued a statement concerning the possible adverse influence of blueish street lighting on the sleep-wake cycle of city-dwellers. Industry critics claim exposure levels are not high enough to have a noticeable effect.

Advantages

- Efficiency: LEDs emit more lumens per watt than incandescent light bulbs. The efficiency of LED lighting fixtures is not affected by shape and size, unlike fluorescent light bulbs or tubes.

- Color: LEDs can emit light of an intended color without using any color filters as traditional lighting methods need. This is more efficient and can lower initial costs.

- Size: LEDs can be very small (smaller than 2 mm²) and are easily attached to printed circuit boards.

- Warmup time: LEDs light up very quickly. A typical red indicator LED achieves full brightness in under a microsecond. LEDs used in communications devices can have even faster response times.

- Cycling: LEDs are ideal for uses subject to frequent on-off cycling, unlike incandescent and fluorescent lamps that fail faster when cycled often, or high-intensity discharge lamps (HID lamps) that require a long time before restarting.

- Dimming: LEDs can very easily be dimmed either by pulse-width modulation or lowering the forward current. This pulse-width modulation is why LED lights, particularly headlights on cars, when viewed on camera or by some people, seem to flash or flicker. This is a type of stroboscopic effect.

- Cool light: In contrast to most light sources, LEDs radiate very little heat in the form of IR that can cause damage to sensitive objects or fabrics. Wasted energy is dispersed as heat through the base of the LED.

- Slow failure: LEDs mainly fail by dimming over time, rather than the abrupt failure of incandescent bulbs.

- Lifetime: LEDs can have a relatively long useful life. One report estimates 35,000 to 50,000 hours of useful life, though time to complete failure may be shorter or longer. Fluorescent tubes typically are rated at about 10,000 to 25,000 hours, depending partly on the conditions of use, and incandescent light bulbs at 1,000 to 2,000 hours. Several DOE demonstrations have shown that reduced maintenance costs from this extended lifetime, rather than energy savings, is the primary factor in determining the payback period for an LED product.

- Shock resistance: LEDs, being solid-state components, are difficult to damage with external shock, unlike fluorescent and incandescent bulbs, which are fragile.

- Focus: The solid package of the LED can be designed to focus its light. Incandescent and fluorescent sources often require an external reflector to collect light and direct it in a usable manner. For larger LED packages total internal reflection (TIR) lenses are often used to the same effect. However, when large quantities of light are needed many light sources are usually deployed, which are difficult to focus or collimate towards the same target.

Disadvantages

- Temperature dependence: LED performance largely depends on the ambient temperature of the operating environment – or thermal management properties. Overdriving an LED in high ambient temperatures may result in overheating the LED package, eventually leading to device failure. An adequate heat sink is needed to maintain long life. This is especially important in automotive, medical, and military uses where devices must operate over a wide range of temperatures, which require low failure rates. Toshiba has produced LEDs with an operating temperature range of −40 to 100 °C, which suits the LEDs for both indoor and outdoor use in applications such as lamps, ceiling lighting, street lights, and floodlights.

- Voltage sensitivity: LEDs must be supplied with a voltage above their threshold voltage and a current below their rating. Current and lifetime change greatly with a small change in applied voltage. They thus require a current-regulated supply (usually just a series resistor for indicator LEDs).

- Color rendition: Most cool-white LEDs have spectra that differ significantly from a black body radiator like the sun or an incandescent light. The spike at 460 nm and dip at 500 nm can make the color of objects appear differently under cool-white LED illumination than sunlight or incandescent sources, due to metamerism, red surfaces being rendered particularly poorly by typical phosphor-based cool-white LEDs. The same is true with green surfaces.

- Area light source: Single LEDs do not approximate a point source of light giving a spherical light distribution, but rather a lambertian distribution. So, LEDs are difficult to apply to uses needing a spherical light field; however, different fields of light can be manipulated by the application of different optics or "lenses". LEDs cannot provide divergence below a few degrees.

- Light pollution: Because white LEDs emit more short wavelength light than sources such as high-pressure sodium vapor lamps, the increased blue and green sensitivity of scotopic vision means that white LEDs used in outdoor lighting cause substantially more sky glow.

- Efficiency droop: The efficiency of LEDs decreases as the electric current increases. Heating also increases with higher currents, which compromises LED lifetime. These effects put practical limits on the current through an LED in high power applications.

- Impact on insects: LEDs are much more attractive to insects than sodium-vapor lights, so much so that there has been speculative concern about the possibility of disruption to food webs.

- Use in winter conditions: Since they do not give off much heat in comparison to incandescent lights, LED lights used for traffic control can have snow obscuring them, leading to accidents.

- Thermal runaway: Parallel strings of LEDs will not share current evenly due to the manufacturing tolerance in their forward voltage. Running two or more strings from a single current source will likely result in LED failure as the devices warm up. A circuit is required to ensure even distribution of current between parallel strands.

Applications

Daytime running light LEDs of an automobile.

LED uses fall into four major categories:

- Visual signals where light goes more or less directly from the source to the human eye, to convey a message or meaning.

- Illumination where light is reflected from objects to give visual response of these objects.

- Measuring and interacting with processes involving no human vision.

- Narrow band light sensors where LEDs operate in a reverse-bias mode and respond to incident light, instead of emitting light.

Indicators and Signs

The low energy consumption, low maintenance and small size of LEDs has led to uses as status indicators and displays on a variety of equipment and installations. Large-area LED displays are used as stadium displays, dynamic decorative displays, and dynamic message signs on freeways. Thin, lightweight message displays are used at airports and railway stations, and as destination displays for trains, buses, trams, and ferries.

Red and green LED traffic signals.

One-color light is well suited for traffic lights and signals, exit signs, emergency vehicle lighting, ships' navigation lights, and LED-based Christmas lights.

Because of their long life, fast switching times, and visibility in broad daylight due to their high output and focus, LEDs have been used in automotive brake lights and turn signals. The use in brakes improves safety, due to a great reduction in the time needed to light fully, or faster rise time, up to 0.5 second faster than an incandescent bulb. This gives drivers behind more time to react. In a dual intensity circuit (rear markers and brakes) if the LEDs are not pulsed at a fast enough frequency, they can create a phantom array, where ghost images of the LED appear if the eyes quickly scan across the array. White LED headlamps are beginning to appear. Using LEDs has styling advantages because LEDs can form much thinner lights than incandescent lamps with parabolic reflectors.

Due to the relative cheapness of low output LEDs, they are also used in many temporary uses such as glowsticks, throwies, and the photonic textile Lumalive. Artists have also used LEDs for LED art.

Lighting

With the development of high-efficiency and high-power LEDs, it has become possible to use LEDs in lighting and illumination. To encourage the shift to LED lamps and other high-efficiency lighting,in 2008 the US Department of Energy created the L Prize competition. The Philips Lighting North America LED bulb won the first competition on August 3, 2011, after successfully completing 18 months of intensive field, lab, and product testing.

Efficient lighting is needed for sustainable architecture. As of 2011, some LED bulbs provide up to 150 lm/W and even inexpensive low-end models typically exceed 50 lm/W, so that a 6-watt LED could achieve the same results as a standard 40-watt incandescent bulb. Displacing less effective sources such as incandescent lamps and fluorescent lighting reduces electrical energy consumption and its associated emissions.

LEDs are used as street lights and in architectural lighting. The mechanical robustness and long lifetime are used in automotive lighting on cars, motorcycles, and bicycle lights. LED street lights are employed on poles and in parking garages. In 2007, the Italian village of Torraca was the first place to convert its street lighting to LEDs.

Cabin lighting on recent Airbus and Boeing jetliners uses LED lighting. LEDs are also being used in airport and heliport lighting. LED airport fixtures currently include medium-intensity runway lights, runway centerline lights, taxiway centerline and edge lights, guidance signs, and obstruction lighting.

LEDs are also used as a light source for DLP projectors, and to backlight LCD televisions (referred to as LED TVs) and laptop displays. RGB LEDs raise the color gamut by as much as 45%. Screens for TV and computer displays can be made thinner using LEDs for backlighting.

The lower heat radiation compared with incandescent lamps makes LEDs ideal for stage lights, where banks of RGB LEDs can easily change color and decrease heating from traditional stage lighting. In medical lighting, infrared heat radiation can be harmful. In energy conservation, the lower heat output of LEDs also reduces demand on air conditioning systems.

LEDs are small, durable and need little power, so they are used in handheld devices such as flashlights. LED strobe lights or camera flashes operate at a safe, low voltage, instead of the 250+ volts commonly found in xenon flashlamp-based lighting. This is especially useful in cameras on mobile phones, where space is at a premium and bulky voltage-raising circuitry is undesirable.

LEDs are used for infrared illumination in night vision uses including security cameras. A ring of LEDs around a video camera, aimed forward into a retroreflective background, allows chroma keying in video productions.

LED for miners, to increase visibility inside mines.

Los Angeles Vincent Thomas Bridge illuminated with blue LEDs.

LEDs are used in mining operations, as cap lamps to provide light for miners. Research has been done to improve LEDs for mining, to reduce glare and to increase illumination, reducing risk of injury to the miners.

LEDs are increasingly finding uses in medical and educational applications, for example as mood enhancement, and new technologies such as AmBX, exploiting LED versatility. NASA has even sponsored research for the use of LEDs to promote health for astronauts.

Data Communication and other Signalling

Light can be used to transmit data and analog signals. For example, lighting white LEDs can be used in systems assisting people to navigate in closed spaces while searching necessary rooms or objects.

Assistive listening devices in many theaters and similar spaces use arrays of infrared LEDs to send sound to listeners' receivers. Light-emitting diodes (as well as semiconductor lasers) are used to send data over many types of fiber optic cable, from digital audio over TOSLINK cables to the very high bandwidth fiber links that form the Internet backbone. For some time, computers were commonly equipped with IrDA interfaces, which allowed them to send and receive data to nearby machines via infrared.

Because LEDs can cycle on and off millions of times per second, very high data bandwidth can be achieved.

Machine Vision Systems

Machine vision systems often require bright and homogeneous illumination, so features of interest are easier to process. LEDs are often used.

Barcode scanners are the most common example of machine vision applications, and many of those scanners use red LEDs instead of lasers. Optical computer mice use LEDs as a light source for the miniature camera within the mouse.

LEDs are useful for machine vision because they provide a compact, reliable source of light. LED lamps can be turned on and off to suit the needs of the vision system, and the shape of the beam produced can be tailored to match the systems's requirements.

A large LED display behind a disc jockey.

LED digital display that can display four digits and points.

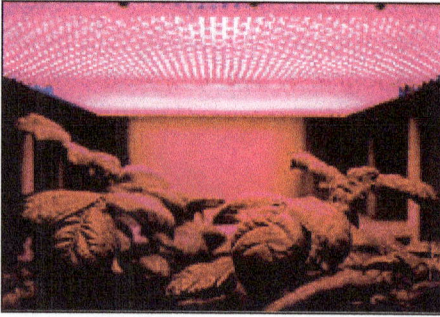

LED panel light source used in an experiment on plant growth. The findings of such experiments may be used to grow food in space on long duration missions.

LED lights reacting dynamically to video feed via AmBX.

Other Applications

The light from LEDs can be modulated very quickly so they are used extensively in optical fiber and free space optics communications. This includes remote controls, such as for television sets, where infrared LEDs are often used. Opto-isolators use an LED combined with a photodiode or phototransistor to provide a signal path with electrical isolation between two circuits. This is especially useful in medical equipment where the signals from a low-voltage sensor circuit (usually battery-powered) in contact with a living organism must be electrically isolated from any possible electrical failure in a recording or monitoring device operating at potentially dangerous voltages. An optoisolator also lets information be transferred between circuits that don't share a common ground potential.

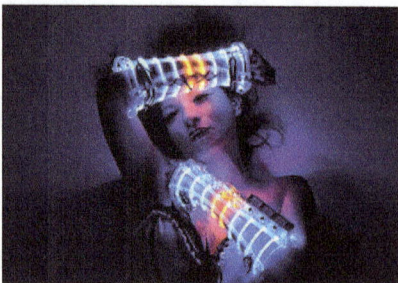

LED costume for stage performers.

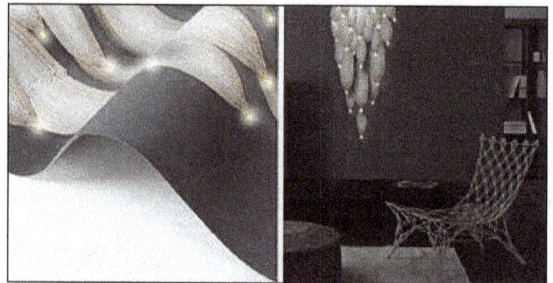

LED wallpaper by Meystyle.

Many sensor systems rely on light as the signal source. LEDs are often ideal as a light source due to the requirements of the sensors. The Nintendo Wii's sensor bar uses infrared LEDs. Pulse oximeters use them for measuring oxygen saturation. Some flatbed scanners use arrays of RGB LEDs rather than the typical cold-cathode fluorescent lamp as the light source. Having independent control of three illuminated colors allows the scanner to calibrate itself for more accurate color balance, and there is no need for warm-up. Further, its sensors only need be monochromatic, since at any one time the page being scanned is only lit by one color of light.

Since LEDs can also be used as photodiodes, they can be used for both photo emission and detection. This could be used, for example, in a touchscreen that registers reflected light from a finger or stylus. Many materials and biological systems are sensitive to, or dependent on, light. Grow lights use LEDs to increase photosynthesis in plants, and bacteria and viruses can be removed from water and other substances using UV LEDs for sterilization.

LEDs have also been used as a medium-quality voltage reference in electronic circuits. The forward voltage drop (about 1.7 V for a red LED or 1.2V for an infrared) can be used instead of a Zener diode in low-voltage regulators. Red LEDs have the flattest I/V curve above the knee. Nitride-based LEDs have a fairly steep I/V curve and are useless for this purpose. Although LED forward voltage is far more current-dependent than a Zener diode, Zener diodes with breakdown voltages below 3 V are not widely available.

The progressive miniaturization of low-voltage lighting technology, such as LEDs and OLEDs, suitable to incorporate into low-thickness materials has fostered experimentation in combining light sources and wall covering surfaces for interior walls in the form of LED wallpaper.

Development

LEDs require optimized efficiency to hinge on ongoing improvements such as phosphor materials and quantum dots.

The process of down-conversion (the method by which materials convert more-energetic photons to different, less energetic colors) also needs improvement. For example, the red phosphors that are used today are thermally sensitive and need to be improved in that aspect so that they do not color shift and experience efficiency drop-off with temperature. Red phosphors could also benefit from a narrower spectral width to emit more lumens and becoming more efficient at converting photons.

In addition, work remains to be done in the realms of current efficiency droop, color shift, system reliability, light distribution, dimming, thermal management, and power supply performance.

Potential Technology

Perovskite LEDs (PLEDs)

A new family of LEDs are based on the semiconductors called perovskites. The ability of perovskite LEDs (PLEDs) to produce light from electrons already rivals those of the best performing OLEDs, and this development has been achieved in less than four years from discovery. They are cost ideal because they can be processed from solution, a low-cost and low-tech method. They optimize efficiency by eliminating non-radiative losses, in other words, elimination of recombination pathways that do not produce photons.

Two-way LEDs

Scientists have discovered a way to create LEDs that can also detect and absorb light. "Nanorods" were created in which quantum dots directly contact two semiconductor materials instead of just one. One semiconductor allows movement of positive charge and one allows movement of negative charge. They can sense light, emit light, and collect energy. The developed nanorod gathers electrons while the quantum dot shell gathers positive charges so the dot emits light, when the voltage is switched the opposite process occurs and the dot absorbs light. So far the only color developed is red.

SCHOTTKY DIODE

Schottky diode is a metal-semiconductor junction diode that has less forward voltage drop than the P-N junction diode and can be used in high-speed switching applications.

In a normal p-n junction diode, a p-type semiconductor and an n-type semiconductor are used to form the p-n junction. When a p-type semiconductor is joined with an n-type semiconductor, a junction is formed between the P-type and N-type semiconductor. This junction is known as P-N junction.

In schottky diode, metals such as aluminum or platinum replace the P-type semiconductor. The schottky diode is named after German physicist Walter H. Schottky.

Schottky diode is also known as schottky barrier diode, surface barrier diode, majority carrier device, hot-electron diode, or hot carrier diode. Schottky diodes are widely used in radio frequency (RF) applications.

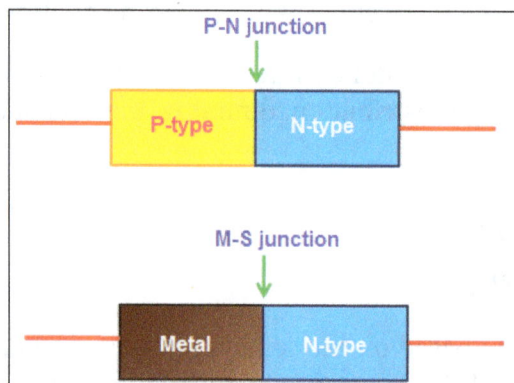

When aluminum or platinum metal is joined with N-type semiconductor, a junction is formed between the metal and N-type semiconductor. This junction is known as a metal-semiconductor junction or M-S junction. A metal-semiconductor junction formed between a metal and n-type semiconductor creates a barrier or depletion layer known as a schottky barrier.

Schottky diode can switch on and off much faster than the p-n junction diode. Also, the schottky diode produces less unwanted noise than p-n junction diode. These two characteristics of the schottky diode make it very useful in high-speed switching power circuits.

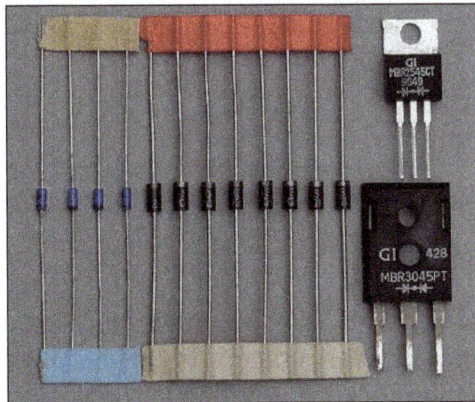

When sufficient voltage is applied to the schottky diode, current starts flowing in the forward direction. Because of this current flow, a small voltage loss occurs across the terminals of the schottky diode. This voltage loss is known as voltage drop.

A silicon diode has a voltage drop of 0.6 to 0.7 volts, while a schottky diode has a voltage drop of 0.2 to 0.3 volts. Voltage loss or voltage drop is the amount of voltage wasted to turn on a diode.

In silicon diode, 0.6 to 0.7 volts is wasted to turn on the diode, whereas in schottky diode, 0.2 to 0.3 volts is wasted to turn on the diode. Therefore, the schottky diode consumes less voltage to turn on.

The voltage needed to turn on the schottky diode is same as that of a germanium diode. But germanium diodes are rarely used because the switching speed of germanium diodes is very low as compared to the schottky diodes.

Symbol of Schottky Diode

The symbol of schottky diode is shown in the below figure. In schottky diode, the metal acts as the anode and n-type semiconductor acts as the cathode.

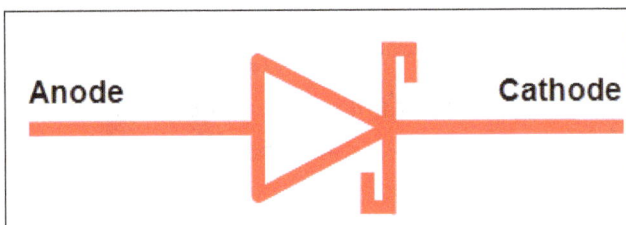

Schottky Diode Symbol.

Metal-semiconductor (M-S) Junction

Metal-semiconductor (M-S) junction is a type of junction formed between a metal and an n-type semiconductor when the metal is joined with the n-type semiconductor. Metal-semiconductor junction is also sometimes referred to as M-S junction.

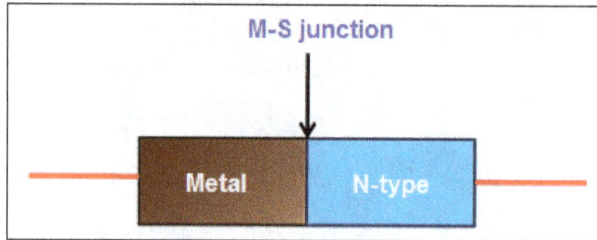

The metal-semiconductor junction can be either non-rectifying or rectifying. The non-rectifying metal-semiconductor junction is called ohmic contact. The rectifying metal-semiconductor junction is called non-ohmic contact.

Schottky Barrier

Schottky barrier is a depletion layer formed at the junction of a metal and n-type semiconductor. In simple words, schottky barrier is the potential energy barrier formed at the metal-semiconductor junction. The electrons have to overcome this potential energy barrier to flow across the diode.

The rectifying metal-semiconductor junction forms a rectifying schottky barrier. This rectifying schottky barrier is used for making a device known as schottky diode. The non-rectifying metal-semiconductor junction forms a non-rectifying schottky barrier.

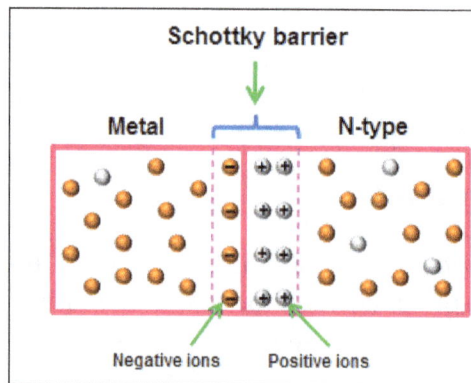

One of the most important characteristics of a schottky barrier is the schottky barrier height. The value of this barrier height depends on the combination of semiconductor and metal.

The schottky barrier height of ohmic contact (non-rectifying barrier) is very low whereas the schottky barrier height of non-ohmic contact (rectifying barrier) is high.

In non-rectifying schottky barrier, the barrier height is not high enough to form a depletion region. So depletion region is negligible or absent in the ohmic contact diode.

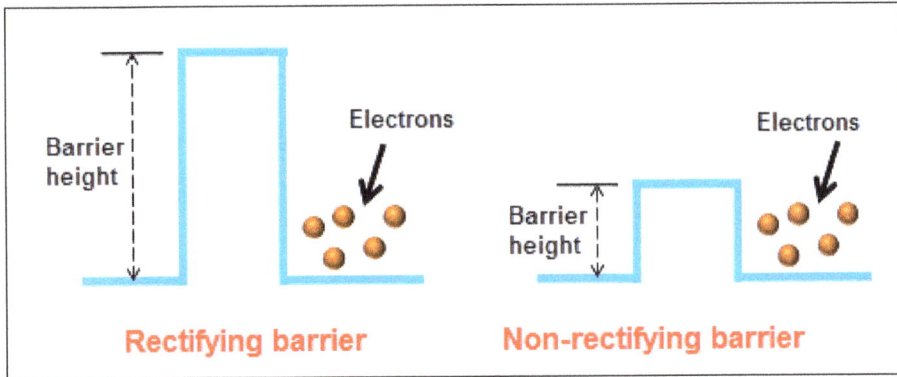

On the other hand, in rectifying schottky barrier, the barrier height is high enough to form a depletion region. So the depletion region is present in the non-ohmic contact diode.

The non-rectifying metal-semiconductor junction (ohmic contact) offers very low resistance to the electric current whereas the rectifying metal-semiconductor junction offers high resistance to the electric current as compared to the ohmic contact.

The rectifying schottky barrier is formed when a metal is in contact with the lightly doped semiconductor, whereas the non-rectifying barrier is formed when a metal is in contact with the heavily doped semiconductor.

The ohmic contact has a linear current-voltage (I-V) curve whereas the non-ohmic contact has a non-linear current-voltage (I-V) curve.

Energy Band Diagram of Schottky Diode

The energy band diagram of the N-type semiconductor and metal is shown in the below figure.

The vacuum level is defined as the energy level of electrons that are outside the material. The work function is defined as the energy required to move an electron from Fermi level (E_F) to vacuum level (E_o).

The work function is different for metal and semiconductor. The work function of a metal is greater than the work function of a semiconductor. Therefore, the electrons in the n-type semiconductor have high potential energy than the electrons in the metal.

The energy levels of the metal and semiconductor are different. The Fermi level at N-type semiconductor side lies above the metal side.

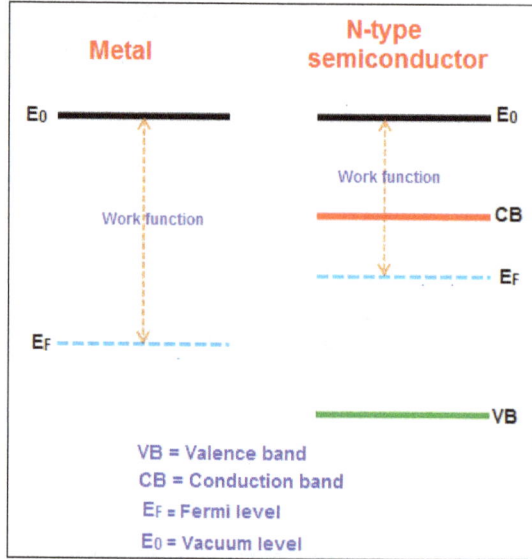

We know that electrons in the higher energy level have more potential energy than the electrons in the lower energy level. So the electrons in the N-type semiconductor have more potential energy than the electrons in the metal.

The energy band diagram of the metal and n-type semiconductor after contact is shown in the below figure.

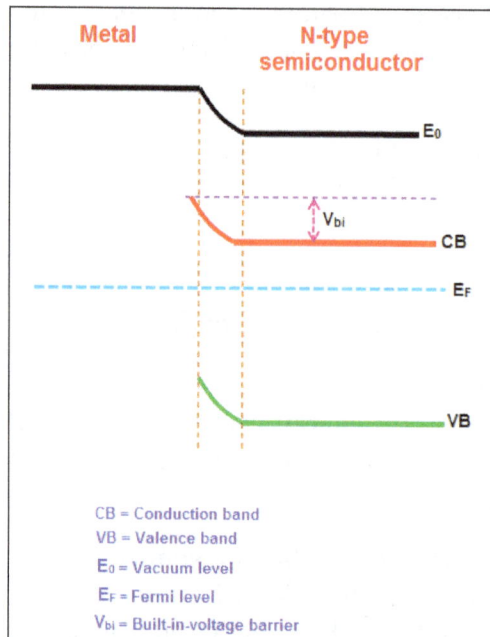

When the metal is joined with the n-type semiconductor, a device is created known as schottky diode. The built-in-voltage (V_{bi}) for schottky diode is given by the difference between the work functions of a metal and n-type semiconductor.

Working of Schottky Diode

Unbiased Schottky Diode

When the metal is joined with the n-type semiconductor, the conduction band electrons (free electrons) in the n-type semiconductor will move from n-type semiconductor to metal to establish an equilibrium state.

We know that when a neutral atom loses an electron it becomes a positive ion similarly when a neutral atom gains an extra electron it becomes a negative ion.

The conduction band electrons or free electrons that are crossing the junction will provide extra electrons to the atoms in the metal. As a result, the atoms at the metal junction gains extra electrons and the atoms at the n-side junction lose electrons.

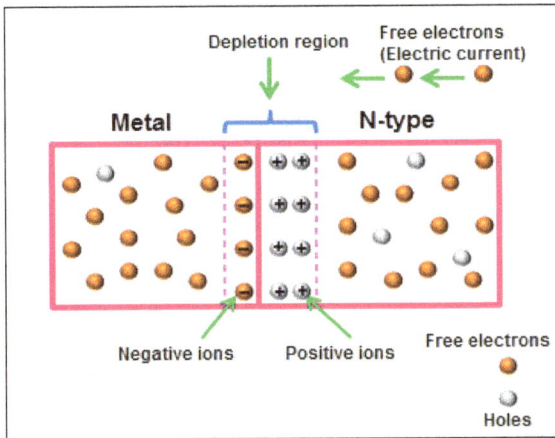

Unbiased schottky diode.

The atoms that lose electrons at the n-side junction will become positive ions whereas the atoms that gain extra electrons at the metal junction will become negative ions. Thus, positive ions are created the n-side junction and negative ions are created at the metal junction. These positive and negative ions are nothing but the depletion region.

Since the metal has a sea of free electrons, the width over which these electrons move into the metal is negligibly thin as compared to the width inside the n-type semiconductor. So the built-in-potential or built-in-voltage is primarily present inside the n-type semiconductor. The built-in-voltage is the barrier seen by the conduction band electrons of the n-type semiconductor when trying to move into the metal.

To overcome this barrier, the free electrons need energy greater than the built-in-voltage. In unbiased schottky diode, only a small number of electrons will flow from n-type semiconductor to metal. The built-in-voltage prevents further electron flow from the semiconductor conduction band into the metal.

The transfer of free electrons from the n-type semiconductor into metal results in energy band bending near the contact.

Forward Biased Schottky Diode

If the positive terminal of the battery is connected to the metal and the negative terminal of the battery is connected to the n-type semiconductor, the schottky diode is said to be forward biased.

When a forward bias voltage is applied to the schottky diode, a large number of free electrons are generated in the n-type semiconductor and metal. However, the free electrons in n-type semiconductor and metal cannot cross the junction unless the applied voltage is greater than 0.2 volts.

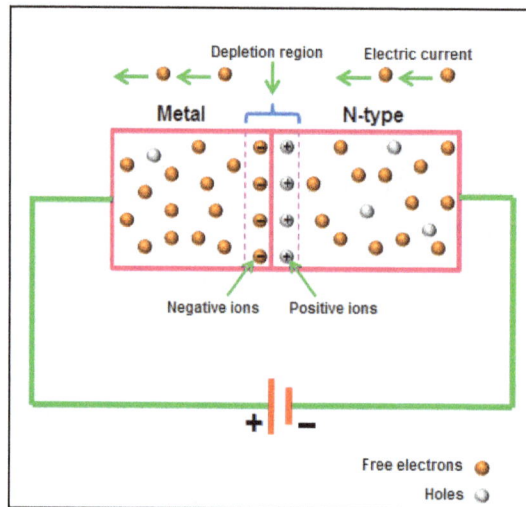

Forward biased schottky diode.

If the applied voltage is greater than 0.2 volts, the free electrons gain enough energy and overcomes the built-in-voltage of the depletion region. As a result, electric current starts flowing through the schottky diode.

If the applied voltage is continuously increased, the depletion region becomes very thin and finally disappears.

Reverse Bias Schottky Diode

If the negative terminal of the battery is connected to the metal and the positive terminal of the battery is connected to the n-type semiconductor, the schottky diode is said to be reverse biased.

When a reverse bias voltage is applied to the schottky diode, the depletion width increases. As a result, the electric current stops flowing. However, a small leakage current flows due to the thermally excited electrons in the metal.

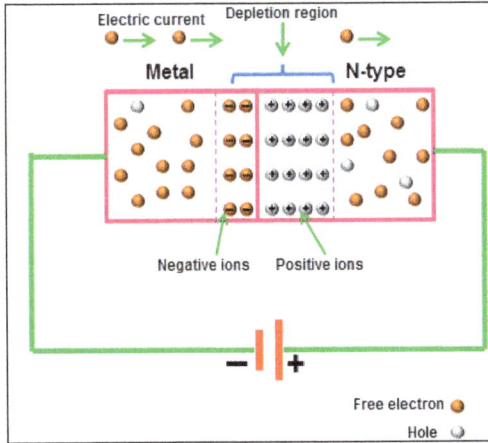

Reverse biased schottky diode.

If the reverse bias voltage is continuously increased, the electric current gradually increases due to the weak barrier.

If the reverse bias voltage is largely increased, a sudden rise in electric current takes place. This sudden rise in electric current causes depletion region to break down which may permanently damage the device.

V-I characteristics of Schottky Diode

The V-I (Voltage-Current) characteristics of schottky diode is shown in the below figure. The vertical line in the below figure represents the current flow in the schottky diode and the horizontal line represents the voltage applied across the schottky diode.

The V-I characteristics of schottky diode is almost similar to the P-N junction diode. However, the forward voltage drop of schottky diode is very low as compared to the P-N junction diode.

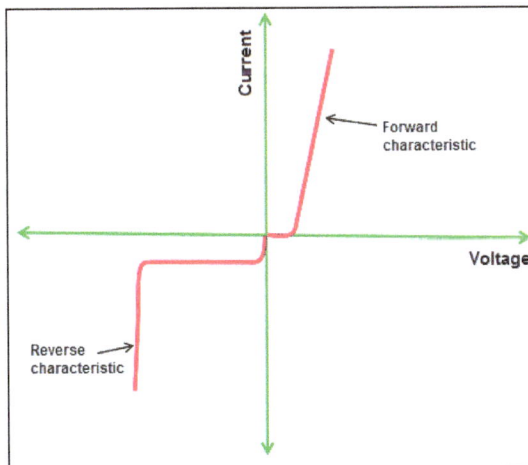

V-I characteristics schottky diode.

The forward voltage drop of schottky diode is 0.2 to 0.3 volts whereas the forward voltage drop of silicon P-N junction diode is 0.6 to 0.7 volts.

If the forward bias voltage is greater than 0.2 or 0.3 volts, electric current starts flowing through the schottky diode.

In schottky diode, the reverse saturation current occurs at a very low voltage as compared to the silicon diode.

Difference between Schottky Diode and P-N Junction Diode

The main difference between schottky diode and p-n junction diode is as follows:

In schottky diode, the free electrons carry most of the electric current. Holes carry negligible electric current. So schottky diode is a unipolar device. In P-N junction diode, both free electrons and holes carry electric current. So P-N junction diode is a bipolar device. The reverse breakdown voltage of a schottky diode is very small as compared to the p-n junction diode.

In schottky diode, the depletion region is absent or negligible, whereas in p-n junction diode the depletion region is present.

The turn-on voltage for a schottky diode is very low as compared to the p-n junction diode.

In schottky diode, electrons are the majority carriers in both metal and semiconductor. In P-N junction diode, electrons are the majority carriers in n-region and holes are the majority carriers in p-region.

Advantages of Schottky Diode

- Low junction capacitance:

We know that capacitance is the ability to store an electric charge. In a P-N junction diode, the depletion region consists of stored charges. So there exists a capacitance. This capacitance is present at the junction of the diode. So it is known as junction capacitance.

In schottky diode, stored charges or depletion region is negligible. So a schottky diode has a very low capacitance.

- Fast reverse recovery time:

The amount of time it takes for a diode to switch from ON state to OFF state is called reverse recovery time.

In order to switch from ON (conducting) state to OFF (non-conducting) state, the stored charges in the depletion region must be first discharged or removed before the diode switch to OFF (non-conducting) state.

The P-N junction diode do not immediately switch from ON state to OFF state because it takes some time to discharge or remove stored charges at the depletion region. However, in schottky diode, the depletion region is negligible. So the schottky diode will immediately switch from ON to OFF state.

- High current density:

We know that the depletion region is negligible in schottky diode. So applying is small voltage is enough to produce large current.

- Low forward voltage drop or low turn on voltage:

The turn on voltage for schottky diode is very small as compared to the P-N junction diode. The turn on voltage for schottky diode is 0.2 to 0.3 volts whereas for P-N junction diode is 0.6 to 0.7 volts. So applying a small voltage is enough to produce electric current in the schottky diode.

- High efficiency.

- Schottky diodes operate at high frequencies.

- Schottky diode produces less unwanted noise than P-N junction diode.

Disadvantages of Schottky Diode

- Large reverse saturation current.

- Schottky diode produces large reverse saturation current than the p-n junction diode.

Applications of Schottky Diodes

- Schottky diodes are used as general-purpose rectifiers.

- Schottky diodes are used in radio frequency (RF) applications.

- Schottky diodes are widely used in power supplies.

- Schottky diodes are used to detect signals.

- Schottky diodes are used in logic circuits.

TUNNEL DIODE

A tunnel diode or Esaki diode is a type of semiconductor diode that has negative resistance due to the quantum mechanical effect called tunneling. It was invented in August

1957 by Leo Esaki, Yuriko Kurose, and Takashi Suzuki when they were working at Tokyo Tsushin Kogyo, now known as Sony. In 1973, Esaki received the Nobel Prize in Physics, jointly with Brian Josephson, for discovering the electron tunneling effect used in these diodes. Robert Noyce independently devised the idea of a tunnel diode while working for William Shockley, but was discouraged from pursuing it. Tunnel diodes were first manufactured by Sony in 1957, followed by General Electric and other companies from about 1960, and are still made in low volume today.

Tunnel diodes have a heavily doped p–n junction that is about 10 nm (100 Å) wide. The heavy doping results in a broken band gap, where conduction band electron states on the n-side are more or less aligned with valence band hole states on the p-side. They are usually made from germanium, but can also be made from gallium arsenide and silicon materials. Their negative differential resistance in part of their operating range allows them to function as oscillators and amplifiers, and in switching circuits using hysteresis. They are also used as frequency converters and detectors. Their low capacitance allows them to function at microwave frequencies, above the range of ordinary diodes and transistors.

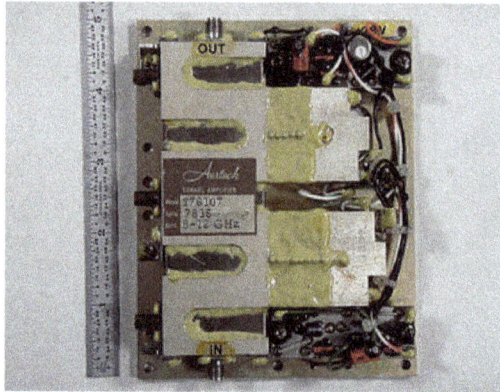

8–12 GHz tunnel diode amplifier, circa 1970.

Tunnel diodes are not widely used due to their low output power; their RF output is limited to several hundred milliwatts due to their small voltage swing. In recent years, however, new devices that use the tunneling mechanism have been developed. The resonant-tunneling diode (RTD) has achieved some of the highest frequencies of any solid-state oscillator. Another type of tunnel diode is a metal–insulator–metal (MIM) diode, but its present application appears to be limited to research environments due to inherent sensitivities. There is also a metal–insulator–insulator–metal (MIIM) diode, where an additional insulator layer allows "*step tunneling*" for precise diode control.

Forward Bias Operation

Under normal forward bias operation, as voltage begins to increase, electrons at first tunnel through the very narrow p–n junction barrier and fill electron states in the conduction band on the n-side which become aligned with empty valence band hole states on the p-side of the p-n junction. As voltage increases further, these states become increasingly misaligned, and the current drops. This is called negative differential resistance because current decreases with increasing voltage. As voltage increases, the diode begins to operate as a normal diode, where electrons travel by conduction across the p–n junction, and no longer by tunneling through the p–n junction barrier. The most important operating region for a tunnel diode is the negative resistance region. Its graph is different from normal p–n junction diode.

Reverse Bias Operation

When used in the reverse direction, tunnel diodes are called back diodes (or backward diodes) and can act as fast rectifiers with zero offset voltage and extreme linearity for power signals (they have an accurate square law characteristic in the reverse direction). Under reverse bias, filled states on the p-side become increasingly aligned with empty states on the n-side, and electrons now tunnel through the p–n junction barrier in reverse direction.

Technical Comparisons

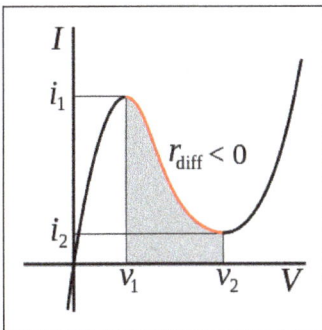

IV curve similar to a tunnel diode characteristic curve. It has negative differential resistance in the shaded voltage region, between v_1 and v_2.

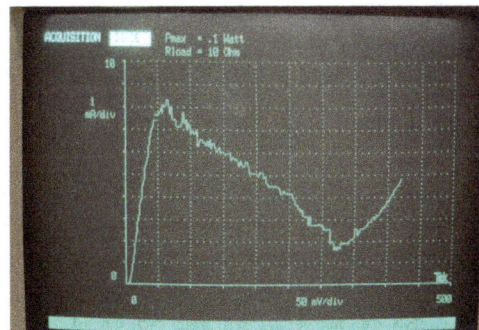

I-V curve of 10 mA germanium tunnel diode, taken on a Tektronix model 571 curve tracer.

In a conventional semiconductor diode, conduction takes place while the p–n junction is forward biased and blocks current flow when the junction is reverse biased. This occurs up to a point known as the "reverse breakdown voltage" at which point conduction begins (often accompanied by destruction of the device). In the tunnel diode, the dopant concentrations in the p and n layers are increased to a level such that the reverse breakdown voltage becomes zero and the diode conducts in the reverse direction. However, when forward-biased, an effect occurs called quantum mechanical tunneling which gives rise to a region in its voltage-current behavior where an *increase* in forward voltage is accompanied by a *decrease* in forward current. This negative resistance region can be exploited in a solid state version of the dynatron oscillator which normally uses a tetrode thermionic valve (vacuum tube).

Applications

The tunnel diode showed great promise as an oscillator and high-frequency threshold (trigger) device since it operated at frequencies far greater than the tetrode could: well into the microwave bands. Applications of tunnel diodes included local oscillators for UHF television tuners, trigger circuits in oscilloscopes, high-speed counter circuits, and very fast-rise time pulse generator circuits. In 1977, the Intelsat V satellite receiver used a microstrip tunnel diode amplifier (TDA) front-end in the 14 to 15.5 GHz frequency band. Such amplifiers were considered state-of-the-art, with better performance at high frequencies than any transistor-based front end. The tunnel diode can also be used as a low-noise microwave amplifier. Since its discovery, more conventional semiconductor devices have surpassed its performance using conventional oscillator techniques. For many purposes, a three-terminal device, such as a field-effect transistor, is more flexible than a device with only two terminals. Practical tunnel diodes operate at a few milliamperes and a few tenths of a volt, making them low-power devices. The Gunn diode has similar high frequency capability and can handle more power.

Tunnel diodes are also more resistant to ionizing radiation than other diodes. This makes them well suited to higher radiation environments such as those found in space.

Longevity

Tunnel diodes are notable for their longevity, with devices made in the 1960s still functioning. Writing in *Nature*, Esaki and coauthors state that semiconductor devices in general are extremely stable, and suggest that their shelf life should be "infinite" if kept at room temperature. They go on to report that a small-scale test of 50-year-old devices revealed a "gratifying confirmation of the diode's longevity". As noticed on some samples of Esaki diodes, the gold plated iron pins can in fact corrode and short out to the case. This can usually be diagnosed and treated with simple peroxide/vinegar technique normally used for repairing phone PCBs and the diode inside normally still works.

These components are susceptible to damage by overheating, and thus special care is needed when soldering them. Surplus Russian units are also reliable and often can be purchased for a few pence despite original cost being in the £30–50 range. The units typically sold are GaAs based and have a I_{pk}/I_v ratio of 5:1 at around 1–20 mA I_{pk}, and so should be protected against overcurrent.

LASER DIODE

A laser diode, (LD), injection laser diode (ILD), or diode laser is a semiconductor device similar to a light-emitting diode in which a laser beam is created at the diode's junction. Laser diodes can directly convert electrical energy into light. Driven by voltage, the doped p-n-transition allows for recombination of an electron with a hole. Due to the drop of the electron from a higher energy level to a lower one, radiation, in the form of an emitted photon is generated. This is spontaneous emission. Stimulated emission can be produced when the process is continued and further generate light with the same phase, coherence and wavelength.

A packaged laser diode shown with a penny for scale.

The laser diode chip is removed from the above package and placed on the eye of a needle for scale.

A laser diode with the case cut away. The laser diode chip is the small black chip at the front; a photodiode at the back is used to control output power.

The choice of the semiconductor material determines the wavelength of the emitted beam, which in today's laser diodes range from infra-red to the UV spectrum. Laser diodes are the most common type of lasers produced, with a wide range of uses that include fiber optic communications, barcode readers, laser pointers, CD/DVD/Blu-ray disc reading/recording, laser printing, laser scanning and light beam illumination. With the use of a phosphor like that found on white LEDs, Laser diodes can be used for general illumination.

Theory of Operation of Simple Diode

A laser diode is electrically a PIN diode. The active region of the laser diode is in the intrinsic (I) region, and the carriers (electrons and holes) are pumped into that region from the N and P regions respectively. While initial diode laser research was conducted on simple P-N diodes, all modern lasers use the double-hetero-structure implementation, where the carriers and the photons are confined in order to maximize their chances for recombination and light generation. Unlike a regular diode, the goal for a laser diode is to recombine all carriers in the I region, and produce light. Thus, laser diodes are fabricated using direct band-gap semiconductors. The laser diode epitaxial structure is grown using one of the crystal growth techniques, usually starting from an N doped substrate, and growing the I doped active layer, followed by the P doped cladding, and a contact layer. The active layer most often consists of quantum wells, which provide lower threshold current and higher efficiency.

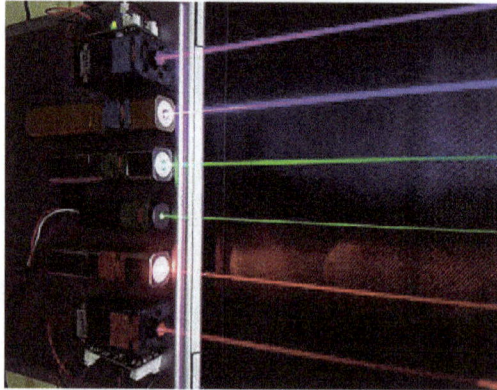

Semi-conductor lasers (660 nm, 635 nm, 532 nm, 520 nm, 445 nm, 405 nm).

Electrical and Optical Pumping

Laser diodes form a subset of the larger classification of semiconductor p-n junction diodes. Forward electrical bias across the laser diode causes the two species of charge carrier – holes and electrons – to be "injected" from opposite sides of the p-n junction into the depletion region. Holes are injected from the p-doped, and electrons from the n-doped, semiconductor. (A depletion region, devoid of any charge carriers, forms as a result of the difference in electrical potential between n- and p-type semiconductors wherever they are in physical contact.) Due to the use of charge injection in powering most diode lasers, this class of lasers is sometimes termed "injection lasers," or "injection laser diode" (ILD). As diode lasers are semiconductor devices, they may also be classified as semiconductor lasers. Either designation distinguishes diode lasers from solid-state lasers.

Another method of powering some diode lasers is the use of optical pumping. Optically pumped semiconductor lasers (OPSL) use a III-V semiconductor chip as the gain medium, and another laser (often another diode laser) as the pump source. OPSL offer several advantages over ILDs, particularly in wavelength selection and lack of interference from internal electrode structures. A further advantage of OPSLs is invariance of the beam parameters - divergence, shape, and pointing - as pump power (and hence output power) is varied, even over a 10:1 output power ratio.

Generation of Spontaneous Emission

When an electron and a hole are present in the same region, they may recombine or "annihilate" producing a spontaneous emission — i.e., the electron may re-occupy the energy state of the hole, emitting a photon with energy equal to the difference between the electron's original state and hole's state. (In a conventional semiconductor junction diode, the energy released from the recombination of electrons and holes is carried away as phonons, i.e., lattice vibrations, rather than as photons.) Spontaneous emission below the lasing threshold produces similar properties to an LED. Spontaneous emission is necessary to initiate laser oscillation, but it is one among several sources of inefficiency once the laser is oscillating.

Direct and Indirect Bandgap Semiconductors

The difference between the photon-emitting semiconductor laser and a conventional phonon-emitting (non-light-emitting) semiconductor junction diode lies in the type of semiconductor used, one whose physical and atomic structure confers the possibility for photon emission. These photon-emitting semiconductors are the so-called "direct bandgap" semiconductors. The properties of silicon and germanium, which are single-element semiconductors, have bandgaps that do not align in the way needed to allow photon emission and are not considered "direct." Other materials, the so-called compound semiconductors, have virtually identical crystalline structures as silicon or germanium but use alternating arrangements of two different atomic species in a checkerboard-like pattern to break the symmetry. The transition between the materials in the alternating pattern creates the critical "direct bandgap" property. Gallium arsenide, indium phosphide, gallium antimonide, and gallium nitride are all examples of compound semiconductor materials that can be used to create junction diodes that emit light.

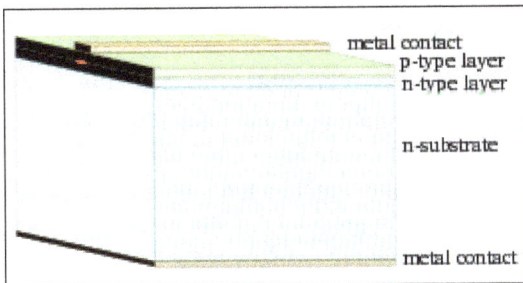

Diagram of a simple laser diode, such as shown above; not to scale.

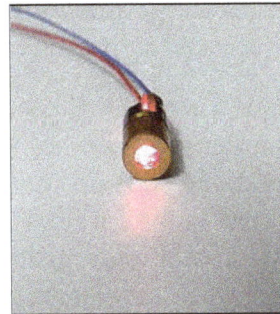

A simple and low power metal enclosed laser diode.

Generation of Stimulated Emission

In the absence of stimulated emission (e.g., lasing) conditions, electrons and holes may coexist in proximity to one another, without recombining, for a certain time, termed the "upper-state lifetime" or "recombination time" (about a nanosecond for typical diode laser materials), before they recombine. A nearby photon with energy equal to the recombination energy can cause recombination by stimulated emission. This generates another photon of the same frequency, polarization, and phase, travelling in the same direction as the first photon. This means that stimulated emission will cause gain in an optical wave (of the correct wavelength) in the injection region, and the gain increases as the number of electrons and holes injected across the junction increases. The spontaneous and stimulated emission processes are vastly more efficient in direct bandgap semiconductors than in indirect bandgap semiconductors; therefore silicon is not a common material for laser diodes.

Optical Cavity and Laser Modes

As in other lasers, the gain region is surrounded with an optical cavity to form a laser. In the simplest form of laser diode, an optical waveguide is made on that crystal's surface, such that the light is confined to a relatively narrow line. The two ends of the crystal are cleaved to form perfectly smooth, parallel edges, forming a Fabry–Pérot resonator. Photons emitted into a mode of the waveguide will travel along the waveguide and be reflected several times from each end face before they exit. As a light wave passes through the cavity, it is amplified by stimulated emission, but light is also lost due to absorption and by incomplete reflection from the end facets. Finally, if there is more amplification than loss, the diode begins to "lase".

Some important properties of laser diodes are determined by the geometry of the optical cavity. Generally, the light is contained within a very thin layer, and the structure supports only a single optical mode in the direction perpendicular to the layers. In the transverse direction, if the waveguide is wide compared to the wavelength of light, then the waveguide can support multiple transverse optical modes, and the laser is known as "multi-mode". These transversely multi-mode lasers are adequate in cases where one needs a very large amount of power, but not a small diffraction-limited beam; for example in printing, activating chemicals, or pumping other types of lasers.

In applications where a small focused beam is needed, the waveguide must be made narrow, on the order of the optical wavelength. This way, only a single transverse mode is supported and one ends up with a diffraction-limited beam. Such single spatial mode devices are used for optical storage, laser pointers, and fiber optics. Note that these lasers may still support multiple longitudinal modes, and thus can lase at multiple wavelengths simultaneously. The wavelength emitted is a function of the band-gap of the semiconductor material and the modes of the optical cavity. In general, the maximum gain will occur for photons with energy slightly above the band-gap energy, and the modes nearest the peak of the gain curve will lase most strongly. The width of the gain curve will determine the number of additional "side modes" that may also lase, depending on the operating conditions. Single spatial mode lasers that can support multiple longitudinal modes are called Fabry Perot (FP) lasers. An FP laser will lase at multiple cavity modes within the gain bandwidth of the lasing medium. The number of lasing modes in an FP laser is usually unstable, and can fluctuate due to changes in current or temperature.

Single spatial mode diode lasers can be designed so as to operate on a single longitudinal mode. These single frequency diode lasers exhibit a high degree of stability, and are used in spectroscopy and metrology, and as frequency references. Single frequency diode lasers are classed as either distributed feedback (DFB) lasers or distributed Bragg reflector (DBR) lasers.

Formation of Laser Beam

Due to diffraction, the beam diverges (expands) rapidly after leaving the chip, typically at 30 degrees vertically by 10 degrees laterally. A lens must be used in order to form a collimated beam like that produced by a laser pointer. If a circular beam is required, cylindrical lenses and other optics are used. For single spatial mode lasers, using symmetrical lenses, the collimated beam ends up being elliptical in shape, due to the difference in the vertical and lateral divergences. This is easily observable with a red laser pointer.

The simple diode has been heavily modified in recent years to accommodate modern technology, resulting in a variety of types of laser diodes.

Types

The simple laser diode structure is extremely inefficient. Such devices require so much power that they can only achieve pulsed operation without damage. Although historically important and easy to explain, such devices are not practical.

Double Heterostructure Lasers

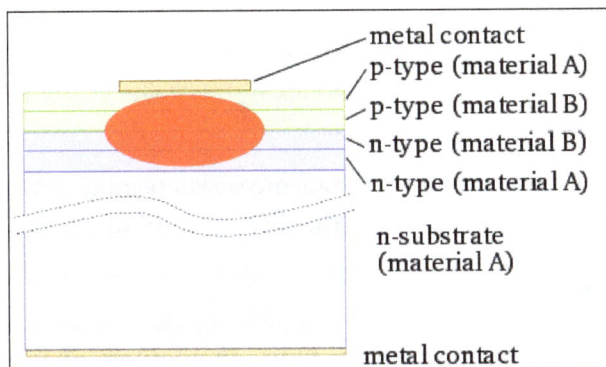

Diagram of front view of a double heterostructure laser diode; not to scale.

In these devices, a layer of low bandgap material is sandwiched between two high bandgap layers. One commonly-used pair of materials is gallium arsenide (GaAs) with aluminium gallium arsenide ($Al_xGa_{(1-x)}As$). Each of the junctions between different bandgap materials is called a *heterostructure*, hence the name "double heterostructure laser" or *DH* laser. The kind of laser diode described in the first part may be referred to as a *homojunction* laser, for contrast with these more popular devices.

The advantage of a DH laser is that the region where free electrons and holes exist simultaneously—the active region—is confined to the thin middle layer. This means that many more of the electron-hole pairs can contribute to amplification—not so many are left out in the poorly amplifying periphery. In addition, light is reflected within the heterojunction; hence, the light is confined to the region where the amplification takes place.

Quantum Well Lasers

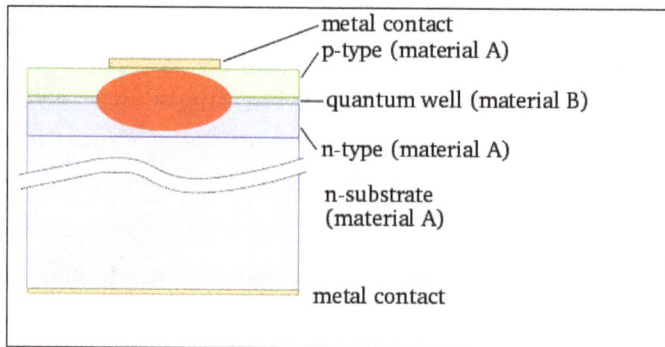

Diagram of front view of a simple quantum well laser diode; not to scale.

If the middle layer is made thin enough, it acts as a quantum well. This means that the vertical variation of the electron's wavefunction, and thus a component of its energy, is quantized. The efficiency of a quantum well laser is greater than that of a bulk laser because the density of states function of electrons in the quantum well system has an abrupt edge that concentrates electrons in energy states that contribute to laser action.

Lasers containing more than one quantum well layer are known as *multiple quantum well* lasers. Multiple quantum wells improve the overlap of the gain region with the optical waveguide mode.

Further improvements in the laser efficiency have also been demonstrated by reducing the quantum well layer to a quantum wire or to a "sea" of quantum dots.

Quantum Cascade Lasers

In a quantum cascade laser, the difference between quantum well energy levels is used for the laser transition instead of the bandgap. This enables laser action at relatively long wavelengths, which can be tuned simply by altering the thickness of the layer. They are heterojunction lasers.

Interband Cascade Lasers

A Interband cascade laser (ICL) is a type of laser diode that can produce coherent radiation over a large part of the mid-infrared region of the electromagnetic spectrum.

Separate Confinement Heterostructure Lasers

The problem with the simple quantum well diode is that the thin layer is simply too small to effectively confine the light. To compensate, another two layers are added on, outside the first three. These layers have a lower refractive index than the centre layers, and hence confine the light effectively. Such a design is called a separate confinement heterostructure (SCH) laser diode.

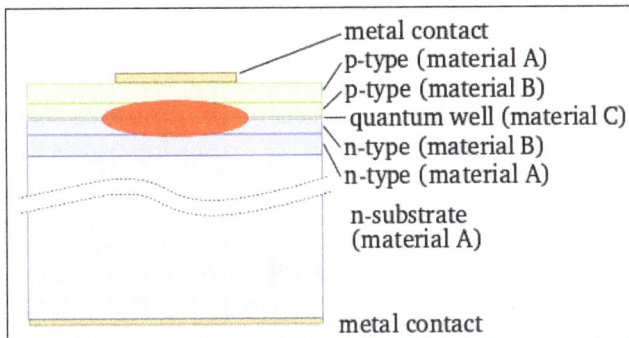

Diagram of front view of a separate confinement heterostructure quantum well laser diode; not to scale.

Almost all commercial laser diodes since the 1990s have been SCH quantum well diodes.

Distributed Bragg Reflector lasers

A distributed Bragg reflector laser (DBR) is a type of single frequency laser diode. It is characterized by an optical cavity consisting of an electrically or optically pumped gain region between two mirrors to provide feedback. One of the mirrors is a broadband reflector and the other mirror is wavelength selective so that gain is favored on a single longitudinal mode, resulting in lasing at a single resonant frequency. The broadband mirror is usually coated with a low reflectivity coating to allow emission. The wavelength selective mirror is a periodically structured diffraction grating with high reflectivity. The diffraction grating is within a non-pumped, or passive region of the cavity. A DBR laser is a monolithic single chip device with the grating etched into the semiconductor. DBR lasers can be edge emitting lasers or VCSELs. Alternative hybrid architectures that share the same topology include extended cavity diode lasers and volume Bragg grating lasers, but these are not properly called DBR lasers.

Distributed Feedback Lasers

A distributed feedback laser (DFB) is a type of single frequency laser diode. DFBs are the most common transmitter type in DWDM-systems. To stabilize the lasing wavelength, a diffraction grating is etched close to the p-n junction of the diode. This grating acts like an optical filter, causing a single wavelength to be fed back to the gain region and lase. Since the grating provides the feedback that is required for lasing, reflection from the facets is not required. Thus, at least one facet of a DFB is anti-reflection coated. The DFB laser has a stable wavelength that is set during manufacturing by the pitch of the grating, and can only be tuned slightly with temperature. DFB lasers are widely used in optical communication applications where a precise and stable wavelength is critical.

The threshold current of this DFB laser, based on its static characteristic, is around 11 mA. The appropriate bias current in a linear regime could be taken in the middle of the static characteristic (50 mA).Several techniques have been proposed in order to enhance the single-mode operation in these kinds of lasers by inserting a one-phase-shift (1PS) or multiple-phase-shift (MPS) in the uniform Bragg grating. However, multiple-phase-shift DFB lasers represent the optimal solution because they have the combination of higher side-mode suppression ratio and reduced spatial hole-burning.

Vertical-cavity surface-emitting Laser

Vertical-cavity surface-emitting lasers (VCSELs) have the optical cavity axis along the direction of current flow rather than perpendicular to the current flow as in conventional laser diodes. The active region length is very short compared with the lateral dimensions so that the radiation emerges from the surface of the cavity rather than from its edge as shown in the figure. The reflectors at the ends of the cavity are dielectric mirrors made from alternating high and low refractive index quarter-wave thick multilayer.

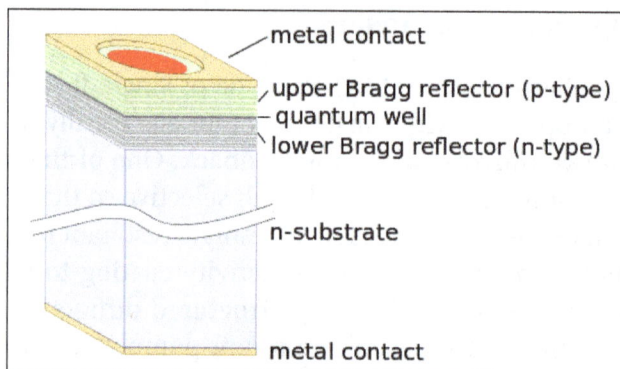

Diagram of a simple VCSEL structure; not to scale.

Such dielectric mirrors provide a high degree of wavelength-selective reflectance at the required free surface wavelength λ if the thicknesses of alternating layers d_1 and d_2

with refractive indices n_1 and n_2 are such that $n_1d_1 + n_2d_2 = \lambda/2$ which then leads to the constructive interference of all partially reflected waves at the interfaces. But there is a disadvantage: because of the high mirror reflectivities, VCSELs have lower output powers when compared to edge-emitting lasers.

There are several advantages to producing VCSELs when compared with the production process of edge-emitting lasers. Edge-emitters cannot be tested until the end of the production process. If the edge-emitter does not work, whether due to bad contacts or poor material growth quality, the production time and the processing materials have been wasted.

Additionally, because VCSELs emit the beam perpendicular to the active region of the laser as opposed to parallel as with an edge emitter, tens of thousands of VCSELs can be processed simultaneously on a three-inch gallium arsenide wafer. Furthermore, even though the VCSEL production process is more labor- and material-intensive, the yield can be controlled to a more predictable outcome. However, they normally show a lower power output level.

Vertical-external-cavity Surface-emitting-laser

Vertical external-cavity surface-emitting lasers, or VECSELs, are similar to VCSELs. In VCSELs, the mirrors are typically grown epitaxially as part of the diode structure, or grown separately and bonded directly to the semiconductor containing the active region. VECSELs are distinguished by a construction in which one of the two mirrors is external to the diode structure. As a result, the cavity includes a free-space region. A typical distance from the diode to the external mirror would be 1 cm.

One of the most interesting features of any VECSEL is the small thickness of the semiconductor gain region in the direction of propagation, less than 100 nm. In contrast, a conventional in-plane semiconductor laser entails light propagation over distances of from 250 μm upward to 2 mm or longer. The significance of the short propagation distance is that it causes the effect of "antiguiding" nonlinearities in the diode laser gain region to be minimized. The result is a large-cross-section single-mode optical beam which is not attainable from in-plane ("edge-emitting") diode lasers.

Several workers demonstrated optically pumped VECSELs, and they continue to be developed for many applications including high power sources for use in industrial machining (cutting, punching, etc.) because of their unusually high power and efficiency when pumped by multi-mode diode laser bars. However, because of their lack of p-n junction, optically-pumped VECSELs are not considered "diode lasers", and are classified as semiconductor lasers.

Electrically pumped VECSELs have also been demonstrated. Applications for electrically pumped VECSELs include projection displays, served by frequency doubling of near-IR VECSEL emitters to produce blue and green light.

External-cavity Diode Lasers

External-cavity diode lasers are tunable lasers which use mainly double heterostructures diodes of the $Al_xGa_{(1-x)}As$ type. The first external-cavity diode lasers used intracavity etalons and simple tuning Littrow gratings. Other designs include gratings in grazing-incidence configuration and multiple-prism grating configurations.

Failure Mechanisms

Laser diodes have the same reliability and failure issues as light emitting diodes. In addition they are subject to *catastrophic optical damage* (COD) when operated at higher power.

Many of the advances in reliability of diode lasers in the last 20 years remain proprietary to their developers. The reliability of a laser diode can make or break a product line. Moreover, *reverse engineering* is not always able to reveal the differences between more-reliable and less-reliable diode laser products.

At the edge of a diode laser, where light is emitted, a mirror is traditionally formed by cleaving the semiconductor wafer to form a specularly reflecting plane. This approach is facilitated by the weakness of the crystallographic plane in III-V semiconductor crystals (such as GaAs, InP, GaSb, etc.) compared to other planes. A scratch made at the edge of the wafer and a slight bending force causes a nearly atomically perfect mirror-like cleavage plane to form and propagate in a straight line across the wafer.

But it so happens that the atomic states at the cleavage plane are altered (compared to their bulk properties within the crystal) by the termination of the perfectly periodic lattice at that plane. Surface states at the cleaved plane have energy levels within the (otherwise forbidden) bandgap of the semiconductor.

Essentially, as a result, when light propagates through the cleavage plane and transits to free space from within the semiconductor crystal, a fraction of the light energy is absorbed by the surface states where it is converted to heat by phonon-electron interactions. This heats the cleaved mirror. In addition, the mirror may heat simply because the edge of the diode laser—which is electrically pumped—is in less-than-perfect contact with the mount that provides a path for heat removal. The heating of the mirror causes the bandgap of the semiconductor to shrink in the warmer areas. The bandgap shrinkage brings more electronic band-to-band transitions into alignment with the photon energy causing yet more absorption. This is thermal runaway, a form of positive feedback, and the result can be melting of the facet, known as *catastrophic optical damage*, or COD.

In the 1970s, this problem, which is particularly nettlesome for GaAs-based lasers emitting between 0.630 μm and 1 μm wavelengths (less so for InP-based lasers used for long-haul telecommunications which emit between 1.3 μm and 2 μm), was identified.

Michael Ettenberg, a researcher and later Vice President at RCA Laboratories' David Sarnoff Research Center in Princeton, New Jersey, devised a solution. A thin layer of aluminum oxide was deposited on the facet. If the aluminum oxide thickness is chosen correctly, it functions as an anti-reflective coating, reducing reflection at the surface. This alleviated the heating and COD at the facet.

Since then, various other refinements have been employed. One approach is to create a so-called non-absorbing mirror (NAM) such that the final 10 μm or so before the light emits from the cleaved facet are rendered non-absorbing at the wavelength of interest.

In the very early 1990s, SDL, Inc. began supplying high power diode lasers with good reliability characteristics. CEO Donald Scifres and CTO David Welch presented new reliability performance data at, e.g., SPIE Photonics West conferences of the era. The methods used by SDL to defeat COD were considered to be highly proprietary and were still undisclosed publicly as of June 2006.

In the mid-1990s, IBM Research (Ruschlikon, Switzerland) announced that it had devised its so-called "E2 process" which conferred extraordinary resistance to COD in GaAs-based lasers. This process, too, was undisclosed as of June 2006.

Reliability of high-power diode laser pump bars (used to pump solid-state lasers) remains a difficult problem in a variety of applications, in spite of these proprietary advances. Indeed, the physics of diode laser failure is still being worked out and research on this subject remains active, if proprietary.

Extension of the lifetime of laser diodes is critical to their continued adaptation to a wide variety of applications.

Applications

Laser diodes are numerically the most common laser type, with 2004 sales of approximately 733 million units, as compared to 131,000 of other types of lasers.

Laser diodes can be arrayed to produce very high power outputs, continuous wave
or pulsed. Such arrays may be used to efficiently pump solid-state lasers for
high average power drilling, burning or for inertial confinement fusion.

Telecommunications, Scanning and Spectrometry

Laser diodes find wide use in telecommunication as easily modulated and easily coupled light sources for fiber optics communication. They are used in various measuring instruments, such as rangefinders. Another common use is in barcode readers. Visible lasers, typically red but later also green, are common as laser pointers. Both low and high-power diodes are used extensively in the printing industry both as light sources for scanning (input) of images and for very high-speed and high-resolution printing plate (output) manufacturing. Infrared and red laser diodes are common in CD players, CD-ROMs and DVD technology. Violet lasers are used in HD DVD and Blu-ray technology. Diode lasers have also found many applications in laser absorption spectrometry (LAS) for high-speed, low-cost assessment or monitoring of the concentration of various species in gas phase. High-power laser diodes are used in industrial applications such as heat treating, cladding, seam welding and for pumping other lasers, such as diode-pumped solid-state lasers.

Uses of laser diodes can be categorized in various ways. Most applications could be served by larger solid-state lasers or optical parametric oscillators, but the low cost of mass-produced diode lasers makes them essential for mass-market applications. Diode lasers can be used in a great many fields; since light has many different properties (power, wavelength, spectral and beam quality, polarization, etc.) it is useful to classify applications by these basic properties.

Many applications of diode lasers primarily make use of the "directed energy" property of an optical beam. In this category, one might include the laser printers, barcode readers, image scanning, illuminators, designators, optical data recording, combustion ignition, laser surgery, industrial sorting, industrial machining, and directed energy weaponry. Some of these applications are well-established while others are emerging.

Medical Uses

Laser medicine: medicine and especially dentistry have found many new uses for diode lasers. The shrinking size and cost of the units and their increasing user friendliness makes them very attractive to clinicians for minor soft tissue procedures. Diode wavelengths range from 810 to 1,100 nm, are poorly absorbed by soft tissue, and are not used for cutting or ablation. Soft tissue is not cut by the laser's beam, but is instead cut by contact with a hot charred glass tip. The laser's irradiation is highly absorbed at the distal end of the tip and heats it up to 500 °C to 900 °C. Because the tip is so hot, it can be used to cut soft-tissue and can cause hemostasis through cauterization and carbonization. Diode lasers when used on soft tissue can cause extensive collateral thermal damage to surrounding tissue.

As laser beam light is inherently coherent, certain applications utilize the coherence of laser diodes. These include interferometric distance measurement, holography, coherent communications, and coherent control of chemical reactions.

Laser diodes are used for their "narrow spectral" properties in the areas of range-finding, telecommunications, infra-red countermeasures, spectroscopic sensing, generation of radio-frequency or terahertz waves, atomic clock state preparation, quantum key cryptography, frequency doubling and conversion, water purification (in the UV), and photodynamic therapy (where a particular wavelength of light would cause a substance such as porphyrin to become chemically active as an anti-cancer agent only where the tissue is illuminated by light).

Laser diodes are used for their ability to generate ultra-short pulses of light by the technique known as "mode-locking." Areas of use include clock distribution for high-performance integrated circuits, high-peak-power sources for laser-induced breakdown spectroscopy sensing, arbitrary waveform generation for radio-frequency waves, photonic sampling for analog-to-digital conversion, and optical code-division-multiple-access systems for secure communication.

Common Wavelengths and Uses

Visible Light

- 405 nm – InGaN blue-violet laser, in Blu-ray Disc and HD DVD drives.

- 445–465 nm – InGaN blue laser multimode diode recently introduced (2010) for use in mercury-free high-brightness data projectors.

- 510–525 nm – InGaN Green diodes recently (2010) developed by Nichia and OSRAM for laser projectors.

- 635 nm – AlGaInP better red laser pointers, same power subjectively twice as bright as 650 nm.

- 650–660 nm – GaInP/AlGaInP CD and DVD drives, cheap red laser pointers.

- 670 nm – AlGaInP bar code readers, first diode laser pointers (now obsolete, replaced by brighter 650 nm and 671 nm DPSS).

Infrared

- 760 nm – AlGaInP gas sensing: O_2.

- 785 nm – GaAlAs Compact Disc drives.

- 808 nm – GaAlAs pumps in DPSS Nd:YAG lasers (e.g., in green laser pointers or as arrays in higher-powered lasers).

- 848 nm – laser mice.

- 980 nm – InGaAs pump for optical amplifiers, for Yb:YAG DPSS lasers.

- 1,064 nm – AlGaAs fiber-optic communication, DPSS laser pump frequency.

- 1,310 nm – InGaAsP, InGaAsN fiber-optic communication.
- 1,480 nm – InGaAsP pump for optical amplifiers.
- 1,512 nm – InGaAsP gas sensing: NH_3.
- 1,550 nm – InGaAsP, InGaAsNSb fiber-optic communication.
- 1,625 nm – InGaAsP fiber-optic communication, service channel.
- 1,654 nm – InGaAsP gas sensing: CH_4.
- 1,877 nm – GaInAsSb gas sensing: H_2O.
- 2,004 nm – GaInAsSb gas sensing: CO_2.
- 2,330 nm – GaInAsSb gas sensing: CO_2.
- 2,680 nm – GaInAsSb gas sensing: CO_2.
- 3,030 nm – GaInAsSb gas sensing: C_2H_2.
- 3,330 nm – GaInAsSb gas sensing: CH_4.

VACUUM DIODE

In 1904, Sir John Ambrose Fleming invented the first vacuum tube called vacuum diode. It is also called Fleming valve or thermionic tube. Vacuum diode is an electronic device that allows the electric current in one direction (cathode to anode) and blocks the electric current in another direction (anode to cathode).

Two Electrodes of Vacuum Diode

Vacuum diode is the simplest form of vacuum tube. It consists of two electrodes, a cathode, and an anode or plate. The cathode emits the free electrons. Hence, it is called as emitter. The anode collects the free electrons. Hence, it is called as collector.

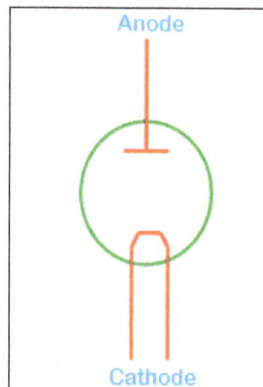

The cathode and anode are enclosed in an empty glass envelope. The anode is a hollow cylinder made of molybdenum or nickel and cathode is a nickel cylinder coated with strontium and barium oxide. The anode surrounds the cathode. In between the cathode and anode an empty space is present, through which the free electrons or electric current flow.

Electrode

Electrode is a conductor through which free electrons or electric current leaves or enters. In vacuum diode, cathode is an electrode or conductor from which the free electrons are emitted into the vacuum. On the other hand, anode is an electrode that collects the free electrons emitted by the cathode. In other words, free electrons leave the cathode and enter into anode.

Electron Emission Depends on the Amount of Heat Applied and the Work Function

The number of free electrons emitted by the cathode is depends on two factors: amount of heat applied and work function. If more amount of heat is applied, more number of free electrons is emitted. Similarly, if less amount of heat is applied, less number of free electrons is emitted.

The minimum amount of energy required to remove the free electrons from the metal is called work function. Metals with low work function will require less amount of heat energy to emit the free electrons. On the other hand, metals with high work function will require large amount of energy to emit the free electrons. Hence, choosing a good material will increase the electron emission efficiency. Most commonly used thermionic emitters include oxide-coated cathode, tungsten, and thoriated tungsten.

Directly and Indirectly Heated Cathode

When the cathode is indirectly or directly heated, free electrons are emitted from it.

In the directly heated cathode, the heat energy is supplied directly to the cathode. Hence, a small amount of heat energy is enough to emit the free electrons from the cathode. When the heat energy is directly supplied to the cathode, large number of free electrons gain sufficient energy and breaks the bonding with the cathode. The free electrons that break the bonding with the cathode are emitted into the vacuum. These emitted free electrons are attracted towards the anode.

In the indirectly heated cathode, no electrical connection is present between the cathode and the heater. Hence, the cathode is not heated directly. The heat energy is supplied to the heater and the heater will transfer its heat energy to the cathode. When the heat energy applied to the cathode is increased to a desired level, the free electrons in the cathode gain sufficient energy and break the bonding with the cathode. The free electrons that break the bonding with the cathode are emitted into the vacuum. These emitted free electrons are attracted towards the anode.

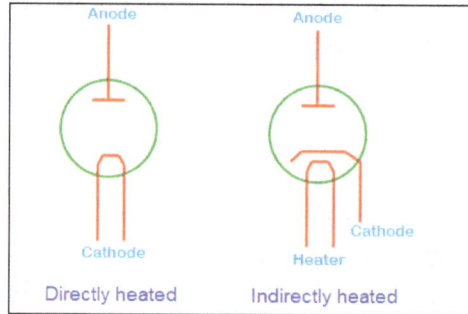

Directly heated   Indirectly heated

Vacuum Diode with Forward Voltage

When the heat is supplied to the heater, it gains heat energy. This heat energy is transferred to the cathode. When the free electrons in the cathode gains sufficient energy, they breaks the bonding with the cathode and jumps into vacuum. The free electrons in the vacuum need sufficient kinetic energy to reach the anode.

If voltage is applied to the vacuum diode, in such a way, that anode is connected to a positive terminal and cathode is connected to a negative terminal (anode is more positive with respect to the cathode), the free electrons in the vacuum gains enough kinetic energy to reach the anode.

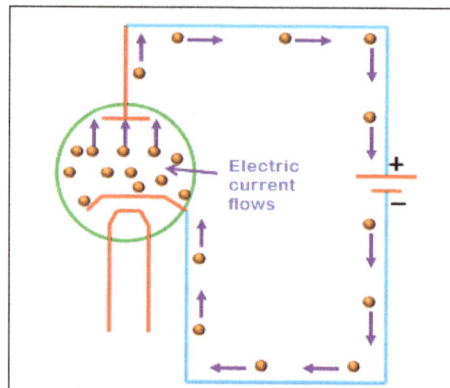

Vacuum diode with forward voltage.

We know that, if two opposite charged particles are placed close to each other they get attracted. In this case, anode is positively charged and free electrons emitted from the cathode are negatively charged. Hence, the free electrons that gain enough kinetic energy will move or attracted towards the anode. These free electrons carry the electric current while moving from cathode to anode.

If the positive voltage applied to plate or anode is increased, the number of free electrons attracted towards the anode is also increased. Thus, the electric current in the vacuum diode increases with increase in the anode or plate voltage.

Vacuum Diode with Reverse Voltage

If voltage is applied to the vacuum diode, in such a way, that anode is connected to the negative terminal and cathode is connected to the positive terminal (anode is more negative with respect to cathode), the free electrons in the vacuum gains enough kinetic energy to reach the anode. However, anode repels the free electrons that try to move towards it.

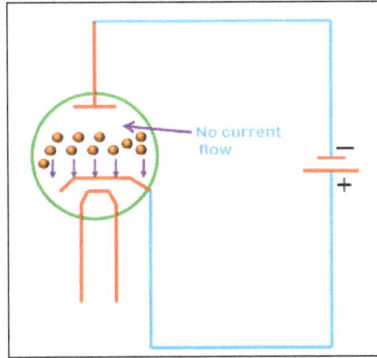

Vacuum diode with reverse voltage.

We know that if two like charged particles are placed close to each other they get repelled. In this case, anode is negatively charged and the free electrons emitted from the cathode are also negatively charged. Hence, the anode repels the free electrons that are emitted by the cathode. Therefore, no electric current flows in the vacuum diode.

Vacuum Diode with Zero Voltage

If no voltage is applied to the vacuum diode, anode or plate acts as neutral. It neither attracts nor repels the free electrons emitted from the cathode. Hence, the free electrons emitted from the cathode do not move or attracted towards the anode.

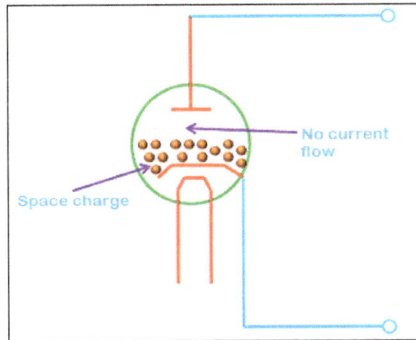

Vacuum diode with zero voltage.

Therefore, no electric current occurs in the vacuum diode. However, the large number of free electrons emitted from the cathode is builds up at one place near the cathode and forms a cloud of free electrons. This cloud of free electrons near the cathode is called space charge.

GUNN DIODE

A Gunn diode, also known as a transferred electron device (TED), is a form of diode, a two-terminal passive semiconductor electronic component, with negative resistance, used in high-frequency electronics. It is based on the "Gunn effect" discovered in 1962 by physicist J. B. Gunn. Its largest use is in electronic oscillators to generate microwaves, in applications such as radar speed guns, microwave relay data link transmitters, and automatic door openers.

A Russian-made Gunn diode.

Its internal construction is unlike other diodes in that it consists only of N-doped semiconductor material, whereas most diodes consist of both P and N-doped regions. It therefore does not conduct in only one direction and cannot rectify alternating current like other diodes, which is why some sources do not use the term diode but prefer TED. In the Gunn diode, three regions exist: two of those are heavily N-doped on each terminal, with a thin layer of lightly n-doped material between. When a voltage is applied to the device, the electrical gradient will be largest across the thin middle layer. If the voltage is increased, the current through the layer will first increase, but eventually, at higher field values, the conductive properties of the middle layer are altered, increasing its resistivity, and causing the current to fall. This means a Gunn diode has a region of negative differential resistance in its current-voltage characteristic curve, in which an increase of applied voltage, causes a decrease in current. This property allows it to amplify, functioning as a radio frequency amplifier, or to become unstable and oscillate when it is biased with a DC voltage.

Gunn Diode Oscillators

The negative differential resistance, combined with the timing properties of the intermediate layer, is responsible for the diode's largest use: in electronic oscillators at microwave frequencies and above. A microwave oscillator can be created simply by applying a DC voltage to bias the device into its negative resistance region. In effect, the negative differential resistance of the diode cancels the positive resistance of the load circuit, thus creating a circuit with zero differential resistance, which will produce spontaneous oscillations. The oscillation frequency is determined partly by the properties of the middle diode layer, but can be tuned by external factors. In practical oscillators, an electronic resonator is usually added to control frequency, in the form of a waveguide, microwave cavity or YIG sphere. The diode is usually mounted inside the cavity. The diode cancels the loss resistance of the resonator, so it produces oscillations at its resonant frequency. The frequency can be tuned mechanically, by adjusting the size of the cavity, or in case of YIG spheres by changing the magnetic field. Gunn diodes are used to build oscillators in the 10 GHz to high (THz) frequency range.

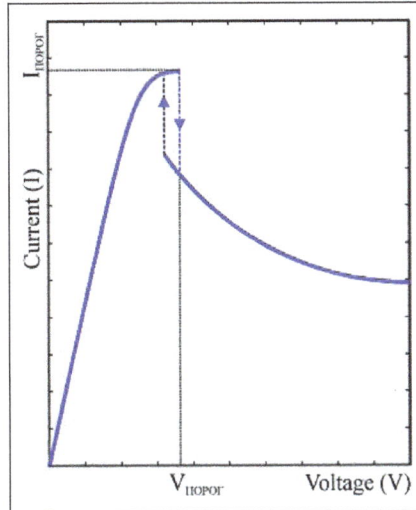

Current-voltage (IV) curve of a Gunn diode. It shows negative resistance above the threshold voltage (V_{nopor}).

Gallium arsenide Gunn diodes are made for frequencies up to 200 GHz, gallium nitride materials can reach up to 3 terahertz.

Working of Gunn Diode Oscillators

The electronic band structure of some semiconductor materials, including gallium arsenide (GaAs), have another energy band or sub-band in addition to the valence and conduction bands which are usually used in semiconductor devices. This third band is at a higher energy than the normal conduction band and is empty until energy is supplied to promote electrons to it. The energy comes from the kinetic energy of ballistic electrons, that is, electrons in the conduction band but moving with sufficient kinetic energy such that they are able to reach the third band.

These electrons either start out below the Fermi level and are given a sufficiently long mean free path to acquire the needed energy by applying a strong electric field, or they are injected by a cathode with the right energy. With forward voltage applied, the Fermi level in the cathode moves into the third band, and reflections of ballistic electrons starting around the Fermi level are minimized by matching the density of states and using the additional interface layers to let the reflected waves interfere destructively.

In GaAs the effective mass of the electrons in the third band is higher than those in the usual conduction band, so the mobility or drift velocity of the electrons in that band is lower. As the forward voltage increases, more and more electrons can reach the third band, causing them to move slower, and current through the device decreases. This creates a region of negative differential resistance in the voltage/current relationship.

When a high enough potential is applied to the diode, the charge carrier density along the cathode becomes unstable, and will develop small segments of low conductivity, with the rest of the cathode having high conductivity. Most of the cathode voltage drop will occur across the segment, so it will have a high electric field. Under the influence of this electric field it will move along the cathode to the anode. It is not possible to balance the population in both bands, so there will always be thin slices of high field strength in a general background of low field strength. So in practice, with a small increase in forward voltage, a low conductivity segment is created at the cathode, resistance increases, the segment moves along the bar to the anode, and when it reaches the anode it is absorbed and a new segment is created at the cathode to keep the total voltage constant. If the voltage is lowered, any existing slice is quenched and resistance decreases again.

The laboratory methods that are used to select materials for the manufacture of Gunn diodes include angle-resolved photoemission spectroscopy.

Applications

Disassembled radar speed gun. The grey assembly attached to the end of the copper-colored horn antenna is the Gunn diode oscillator which generates the microwaves.

Because of their high frequency capability, Gunn diodes are mainly used at microwave frequencies and above. They can produce some of the highest output power of any semiconductor devices at these frequencies. Their most common use is in oscillators, but they are also used in microwave amplifiers to amplify signals. Because the diode is a one-port (two terminal) device, an amplifier circuit must separate the outgoing amplified signal from the incoming input signal to prevent coupling. One common circuit is a *reflection amplifier* which uses a circulator to separate the signals. A bias tee is needed to isolate the bias current from the high frequency oscillations.

Sensors and Measuring Instruments

Gunn diode oscillators are used to generate microwave power for: airborne collision avoidance radar, anti-lock brakes, sensors for monitoring the flow of traffic, car radar detectors, pedestrian safety systems, "distance traveled" recorders, motion detectors, "slow-speed" sensors (to detect pedestrian and traffic movement up to 85 km/h (50 mph)), traffic signal controllers, automatic door openers, automatic traffic gates, process control equipment to monitor throughput, burglar alarms and equipment to detect trespassers, sensors to avoid derailment of trains, remote vibration detectors, rotational speed tachometers, moisture content monitors.

Radio Amateur Use

By virtue of their low voltage operation, Gunn diodes can serve as microwave frequency generators for very low powered (few-milliwatt) microwave transceivers called Gunnplexers. They were first used by British radio amateurs in the late 1970s, and many Gunnplexer designs have been published in journals. They typically consist of an approximately 3 inch waveguide into which the diode is mounted. A low voltage (less than 12 volt) direct current power supply, that can be modulated appropriately, is used to drive the diode. The waveguide is blocked at one end to form a resonant cavity and the other end usually feeds a horn antenna. An additional "mixer diode" is inserted into the waveguide, and it is often connected to a modified FM broadcast receiver to enable listening of other amateur stations. Gunnplexers are most commonly used in the 10 GHz and 24 GHz ham bands and sometimes 22 GHz security alarms are modified as the diode(s) can be put in a slightly detuned cavity with layers of copper or aluminium foil on opposite edges for moving to the licensed amateur band. Typically the mixer diode if intact is reused in its existing waveguide and these parts are well known for being extremely static sensitive. On most commercial units this part is protected with a parallel resistor and other components and a variant is used in some Rb atomic clocks. The mixer diode is useful for lower frequency applications even if the Gunn diode is weakened from use, and some amateur radio enthusiasts have used them in conjunction with an external oscillator or n/2 wavelength Gunn diode for satellite finding and other applications.

Radio Astronomy

Gunn oscillators are used as local oscillators for millimeter-wave and submillimeter-wave radio astronomy receivers. The Gunn diode is mounted in a cavity tuned to resonate at twice the fundamental frequency of the diode. The cavity length is changed by a micrometer adjustment. Gunn oscillators capable of generating over 50 mW over a 50% tuning range (one waveguide band) are available.

The Gunn oscillator frequency is multiplied by a diode frequency multiplier for submillimeter-wave applications.

PHOTODIODE

A photodiode is a semiconductor device that converts light into an electrical current. The current is generated when photons are absorbed in the photodiode. Photodiodes may contain optical filters, built-in lenses, and may have large or small surface areas. Photodiodes usually have a slower response time as their surface area increases. The common, traditional solar cell used to generate electric solar power is a large area photodiode.

Photodiodes are similar to regular semiconductor diodes except that they may be either exposed (to detect vacuum UV or X-rays) or packaged with a window or optical fiber connection to allow light to reach the sensitive part of the device. Many diodes designed for use specially as a photodiode use a PIN junction rather than a p–n junction, to increase the speed of response. A photodiode is designed to operate in reverse bias.

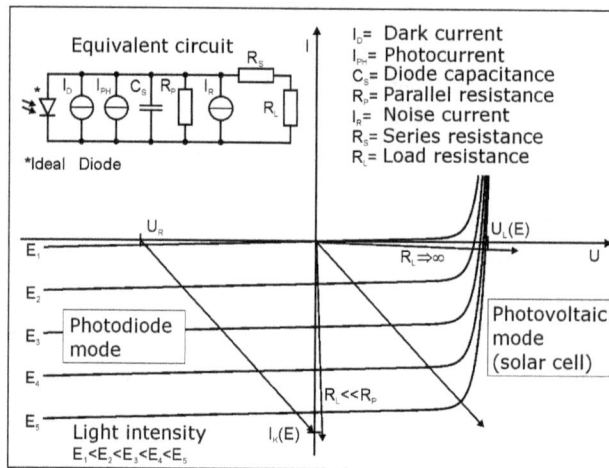

I-V characteristic of a photodiode. The linear load lines represent the response of the external circuit: I=(Applied bias voltage-Diode voltage)/Total resistance. The points of intersection with the curves represent the actual current and voltage for a given bias, resistance and illumination.

Principle of Operation

A photodiode is a p–n junction or PIN structure. When a photon of sufficient energy strikes the diode, it creates an electron–hole pair. This mechanism is also known as the inner photoelectric effect. If the absorption occurs in the junction's depletion region, or one diffusion length away from it, these carriers are swept from the junction by the built-in electric field of the depletion region. Thus holes move toward the anode, and electrons toward the cathode, and a photocurrent is produced. The total current through the photodiode is the sum of the dark current (current that is generated in the absence of light) and the photocurrent, so the dark current must be minimized to maximize the sensitivity of the device.

To first order, for a given spectral distribution, the photocurrent is linearly proportional to the irradiance.

Photovoltaic Mode

When used in zero bias or *photovoltaic mode*, photocurrent flows out of the anode through a short circuit to the cathode. If the circuit is opened or has a load impedance, restricting the photocurrent out of the device, a voltage builds up in the direction that forward biases the diode, that is, anode positive with respect to cathode. If the circuit is open or the impedance is high, a forward current will consume all or some of the photocurrent. This mode exploits the photovoltaic effect, which is the basis for solar cells – a traditional solar cell is just a large area photodiode. For optimum power output, the photovoltaic cell will be operated at a voltage that causes only a small forward current compared to the photocurrent.

Photoconductive Mode

In this mode the diode is reverse biased (with the cathode driven positive with respect to the anode). This reduces the response time because the additional reverse bias increases the width of the depletion layer, which decreases the junction's capacitance and increases the region with an electric field that will cause electrons to be quickly collected. The reverse bias also reduces the dark current without much change in the photocurrent.

Although this mode is faster, the photoconductive mode can exhibit more electronic noise due to dark current or avalanche effects. The leakage current of a good PIN diode is so low (<1 nA) that the Johnson–Nyquist noise of the load resistance in a typical circuit often dominates.

Other Modes of Operation

Avalanche photodiodes are photodiodes with structure optimized for operating with high reverse bias, approaching the reverse breakdown voltage. This allows each *photo-generated* carrier to be multiplied by avalanche breakdown, resulting in internal gain within the photodiode, which increases the effective *responsivity* of the device.

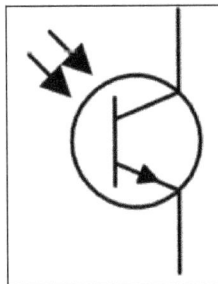

Electronic symbol for a phototransistor.

A phototransistor is a light-sensitive transistor. A common type of phototransistor, called a photobipolar transistor, is in essence a bipolar transistor encased in a transparent case so that light can reach the *base–collector junction*. It was invented by Dr. John N. Shive (more famous for his wave machine) at Bell Labs in 1948, but it was not announced until 1950. The electrons that are generated by photons in the base–collector junction are injected into the base, and this photodiode current is amplified by the transistor's current gain β (or h_{fe}). If the base and collector leads are used and the emitter is left unconnected, the phototransistor becomes a photodiode. While phototransistors have a higher responsivity for light they are not able to detect low levels of light any better than photodiodes. Phototransistors also have significantly longer response times. Field-effect phototransistors, also known as photoFETs, are light-sensitive field-effect transistors. Unlike photobipolar transistors, photoFETs control drain-source current by creating a gate voltage.

A Solaristor is a two-terminal gate-less phototransistor. A compact class of two-terminal phototransistors or solaristors have been demonstrated in 2018 by ICN2 researchers. The novel concept is a two-in-one power source plus transistor device that runs on solar energy by exploiting a memresistive effect in the flow of photogenerated carriers.

Materials

The material used to make a photodiode is critical to defining its properties, because only photons with sufficient energy to excite electrons across the material's bandgap will produce significant photocurrents.

Materials commonly used to produce photodiodes include:

Material	Electromagnetic spectrum wavelength range (nm)
Silicon	190–1100
Germanium	400–1700
Indium gallium arsenide	800–2600
Lead(II) sulfide	<1000–3500
Mercury cadmium telluride	400–14000

Because of their greater bandgap, silicon-based photodiodes generate less noise than germanium-based photodiodes.

Unwanted Photodiode Effects

Any p–n junction, if illuminated, is potentially a photodiode. Semiconductor devices such as diodes, transistors and ICs contain p–n junctions, and will not function correctly if they are illuminated by unwanted electromagnetic radiation (light) of wavelength suitable to produce a photocurrent; this is avoided by encapsulating devices in opaque housings. If these housings are not completely opaque to high-energy radiation (ultraviolet, X-rays, gamma rays), diodes, transistors and ICs can malfunction due to induced photo-currents. Background radiation from the packaging is also significant. Radiation hardening mitigates these effects.

In some cases, the effect is actually wanted, for example to use LEDs as light-sensitive devices or even for energy harvesting, then sometimes called light-emitting and -absorbing diodes (LEADs).

Features

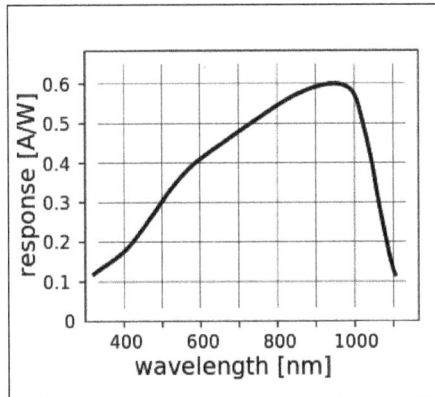

Response of a silicon photo diode vs wavelength of the incident light.

Critical performance parameters of a photodiode include:

Responsivity

The Spectral responsivity is a ratio of the generated photocurrent to incident light power, expressed in A/W when used in photoconductive mode. The wavelength-dependence may also be expressed as a *Quantum efficiency*, or the ratio of the number of photogenerated carriers to incident photons, a unitless quantity.

Dark Current

The current through the photodiode in the absence of light, when it is operated in photoconductive mode. The dark current includes photocurrent generated by background radiation and the saturation current of the semiconductor junction. Dark current must be accounted for by calibration if a photodiode is used to make an accurate optical power measurement, and it is also a source of noise when a photodiode is used in an optical communication system.

Response Time

A photon absorbed by the semiconducting material will generate an electron–hole pair which will in turn start moving in the material under the effect of the electric field and thus generate a current. The finite duration of this current is known as the transit-time spread and can be evaluated by using Ramo's theorem. One can also show with this theorem that the total charge generated in the external circuit is e and not 2e as one might expect by the presence of the two carriers. Indeed, the integral of the current due to both electron and hole over time must be equal to e. The resistance and capacitance of the photodiode and the external circuitry give rise to another response time known as RC time constant $\tau = RC$. This combination of R and C integrates the photoresponse over time and thus lengthens the impulse response of the photodiode. When used in an optical communication system, the response time determines the bandwidth available for signal modulation and thus data transmission.

Noise-equivalent Power

(NEP) The minimum input optical power to generate photocurrent, equal to the rms noise current in a 1 hertz bandwidth. NEP is essentially the minimum detectable power. The related characteristic detectivity (D) is the inverse of NEP, 1/NEP. There is also the specific detectivity (D*) which is the detectivity multiplied by the square root of the area (A) of the photodetector, $(D^* = D\sqrt{A})$ for a 1 Hz bandwidth. The specific detectivity allows different systems to be compared independent of sensor area and system bandwidth; a higher detectivity value indicates a low-noise device or system. Although it is traditional to give (D*) in many catalogues as a measure of the diode's quality, in practice, it is hardly ever the key parameter.

When a photodiode is used in an optical communication system, all these parameters contribute to the *sensitivity* of the optical receiver, which is the minimum input power required for the receiver to achieve a specified *bit error rate*.

Applications

P–n photodiodes are used in similar applications to other photodetectors, such as photoconductors, charge-coupled devices, and photomultiplier tubes. They may be used to generate an output which is dependent upon the illumination (analog; for measurement and the like), or to change the state of circuitry (digital; either for control and switching, or digital signal processing).

Photodiodes are used in consumer electronics devices such as compact disc players, smoke detectors, medical devices and the receivers for infrared remote control devices used to control equipment from televisions to air conditioners. For many applications either photodiodes or photoconductors may be used. Either type of photosensor may be used for light measurement, as in camera light meters, or to respond to light levels, as in switching on street lighting after dark.

Photosensors of all types may be used to respond to incident light, or to a source of light which is part of the same circuit or system. A photodiode is often combined into a single component with an emitter of light, usually a light-emitting diode (LED), either to detect the presence of a mechanical obstruction to the beam (slotted optical switch), or to couple two digital or analog circuits while maintaining extremely high electrical isolation between them, often for safety (optocoupler). The combination of LED and photodiode is also used in many sensor systems to characterize different types of products based on their optical absorbance.

Photodiodes are often used for accurate measurement of light intensity in science and industry. They generally have a more linear response than photoconductors.

They are also widely used in various medical applications, such as detectors for computed tomography (coupled with scintillators), instruments to analyze samples (immunoassay), and pulse oximeters.

PIN diodes are much faster and more sensitive than p–n junction diodes, and hence are often used for optical communications and in lighting regulation.

P–n photodiodes are not used to measure extremely low light intensities. Instead, if high sensitivity is needed, avalanche photodiodes, intensified charge-coupled devices or photomultiplier tubes are used for applications such as astronomy, spectroscopy, night vision equipment and laser rangefinding.

Pinned photodiode is not a PIN photodiode, it has p+/n/p regions in it. It has a shallow P+ implant in N type diffusion layer over a P-type epitaxial substrate layer. It is used in CMOS Active pixel sensor.

Comparison with Photomultipliers

Advantages compared to photomultipliers:

- Excellent linearity of output current as a function of incident light.

- Spectral response from 190 nm to 1100 nm (silicon), longer wavelengths with other semiconductor materials.

- Low noise.

- Ruggedized to mechanical stress.

- Low cost.

- Compact and light weight.

- Long lifetime.

- High quantum efficiency, typically 60–80%.

- No high voltage required.

Disadvantages Compared to Photomultipliers

- Small area.

- No internal gain (except avalanche photodiodes, but their gain is typically 10^2–10^3 compared to 10^5-10^8 for the photomultiplier).

- Much lower overall sensitivity.

- Photon counting only possible with specially designed, usually cooled photodiodes, with special electronic circuits.

- Response time for many designs is slower.

- latent effect.

Photodiode Array

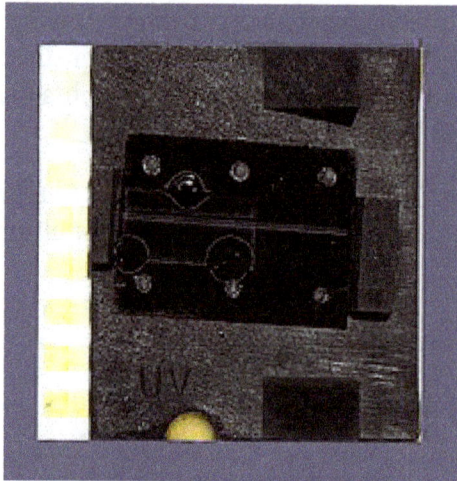

A 2 × 2 cm photodiode array chip with more than 200 diodes.

A one-dimensional array of hundreds or thousands of photodiodes can be used as a position sensor, for example as part of an angle sensor. One advantage of photodiode arrays (PDAs) is that they allow for high speed parallel read out since the driving electronics may not be built in like a traditional CMOS or CCD sensor.

AVALANCHE DIODE

An avalanche diode is a special type of semiconductor device designed to operate in reverse breakdown region. Avalanche diodes are used as relief valves (a type of valve used to control the pressure in a system) to protect electrical systems from excess voltages.

Construction of Avalanche Diode

Avalanche diodes are generally made from silicon or other semiconductor materials. The construction of avalanche diode is similar to zener diode but the doping level in avalanche diode differs from zener diode.

Zener diodes are heavily doped. Therefore, the width of depletion region in zener diode is very thin. Because of this thin depletion layer or region, reverse breakdown occurs at lower voltages in zener diode.

On the other hand, avalanche diodes are lightly doped. Therefore, the width of depletion layer in avalanche diode is very wide compared to the zener diode. Because of this wide depletion region, reverse breakdown occurs at higher voltages in avalanche diode. The breakdown voltage of avalanche diode is carefully set by controlling the doping level during manufacture.

Symbol of Avalanche Diode

The symbol of avalanche and zener diode is same. Avalanche diode consists of two terminals: anode and cathode. The symbol of avalanche diode is shown in below figure.

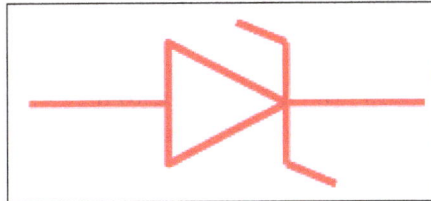

Avalanche diode symbol.

The symbol of avalanche diode is similar to the normal diode but with the bend edges on the vertical bar.

Working of Avalanche Diode

A normal p-n junction diode allows electric current only in forward direction whereas an avalanche diode allows electric current in both forward and reverse directions. However, avalanche diode is specifically designed to operate in reverse biased condition.

Avalanche diode allows electric current in reverse direction when reverse bias voltage exceeds the breakdown voltage. The point or voltage at which electric current increases suddenly is called breakdown voltage.

When the reverse bias voltage applied to the avalanche diode exceeds the breakdown voltage, a junction breakdown occurs. This junction breakdown is called avalanche breakdown.

When forward bias voltage is applied to the avalanche diode, it works like a normal p-n junction diode by allowing electric current through it.

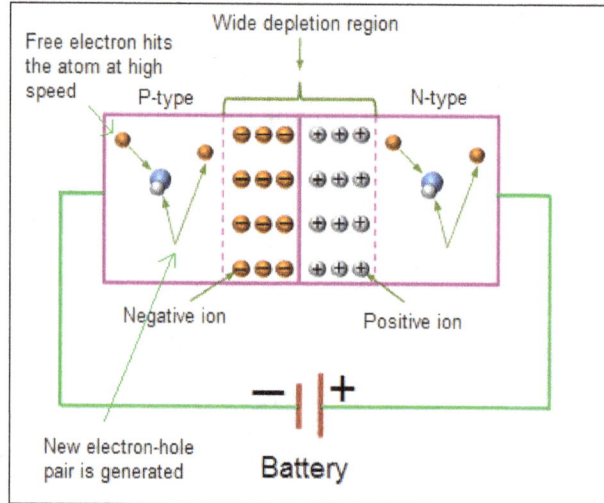

When reverse bias voltage is applied to the avalanche diode, the free electrons (majority carriers) in the n-type semiconductor and the holes (majority carriers) in the p-type semiconductor are moved away from the junction. As a result, the width of depletion region increases. Therefore, the majority carriers will not carry electric current. However, the minority carriers (free electrons in p-type and holes in n-type) experience a repulsive force from external voltage.

As a result, the minority carriers flow from p-type to n-type and n-type to p-type by carrying the electric current. However, electric current carried by minority carriers is very small. This small electric current carried by minority carriers is called reverse leakage current.

If the reverse bias voltage applied to the avalanche diode is further increased, the minority carriers (free electrons or holes) will gain large amount of energy and accelerated to greater velocities. The free electrons moving at high speed will collide with the atoms and transfer their energy to the valence electrons.

The valance electrons which gains enough energy from the high-speed electrons will be detached from the parent atom and become free electrons. These free electrons are again accelerated. When these free electrons again collide with other atoms, they knock off more electrons. Because of this continuous collision with the atoms, a large number of minority carriers (free electrons or holes) are generated. These large numbers of free electrons carry excess current in the diode.

When the reverse voltage applied to the avalanche diode continuously increases, at some point the junction breakdown or avalanche breakdown occurs. At this point, a small increase in voltage will suddenly increases the electric current. This sudden increase of electric current may permanently destroys the normal p-n junction diode. However, avalanche diodes may not be destroyed because they are carefully designed to operate in avalanche breakdown region.

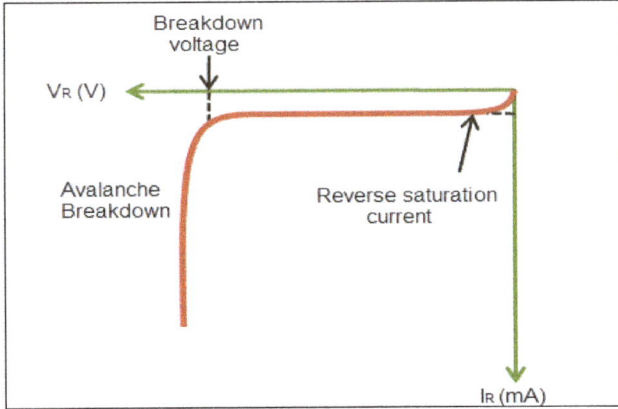

Avalanche breakdown.

The breakdown voltage of the avalanche diode depends on the doping density. Increasing the doping density will decreases the breakdown voltage of the avalanche diode.

Applications of Avalsanche Diodes

- Avalanche diodes can be used as white noise generators.

- Avalanche diodes are used in protecting circuits.

References

- Diode, definition: techtarget.com, Retrieved 9 March, 2019

- Lowe, Doug (2013). "Electronics Components: Diodes". Electronics All-In-One Desk Reference For Dummies. John Wiley & Sons. Retrieved January 4, 2013

- Schottkydiode, semiconductor-diodes, electronic-devices-and-circuits: physics-and-radio-electronics.com, Retrieved 10 April, 2019

- Gribnikov, Z. S., Bashirov, R. R., & Mitin, V. V. (2001). Negative effective mass mechanism of negative differential drift velocity and terahertz generation. IEEE Journal of Selected Topics in Quantum Electronics, 7(4), 630-640

- Vacuumdiode, vacuum-tubes, electronic-devices-and-circuits: physics-and-radio-electronics.com, Retrieved 11 May, 2019

- Rostky, George. "Tunnel diodes: the transistor killers". EE Times. Archived from the original on 7 January 2010. Retrieved 2 October 2009

- Avalanchediode-constructionandworking, semiconductor-diodes, electronic-devices-and-circuits: physics-and-radio-electronics.com, Retrieved 12 June, 2019

4

Thyristor: A Semiconductor Device

Thyristor is a solid-state semiconductor device having four layers of alternating p- and n-type materials. DIAC and TRIAC are a few types of thyristors. The diverse applications of these types of thyristors have been thoroughly discussed in this chapter.

Thyristor is a multi-layer semiconductor device, hence the "silicon" part of its name. It requires a gate signal to turn it "ON", the "controlled" part of the name and once "ON" it behaves like a rectifying diode, the "rectifier" part of the name. In fact the circuit symbol for the *thyristor* suggests that this device acts like a controlled rectifying diode.

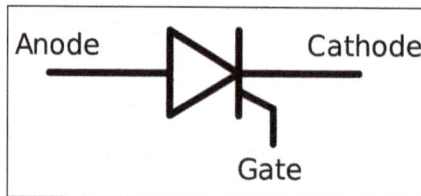

Thyristor Symbol.

However, unlike the junction diode which is a two layer (P-N) semiconductor device, or the commonly used bipolar transistor which is a three layer (P-N-P, or N-P-N) switching device, the Thyristor is a four layer (P-N-P-N) semiconductor device that contains three PN junctions in series.

Like the diode, the Thyristor is a unidirectional device, that is it will only conduct current in one direction only, but unlike a diode, the thyristor can be made to operate as either an open-circuit switch or as a rectifying diode depending upon how the thyristors gate is triggered. In other words, thyristors can operate only in the switching mode and cannot be used for amplification.

The silicon controlled rectifier SCR, is one of several power semiconductor devices along with Triacs (Triode AC's), Diacs (Diode AC's) and UJT's (Unijunction Transistor) that are all capable of acting like very fast solid state AC switches for controlling large AC voltages and currents. So for the Electronics student this makes these very handy solid state devices for controlling AC motors, lamps and for phase control.

The thyristor is a three-terminal device labelled: "Anode", "Cathode" and "Gate" and consisting of three PN junctions which can be switched "ON" and "OFF" at an extremely fast rate, or it can be switched "ON" for variable lengths of time during half cycles to deliver a selected amount of power to a load. The operation of the thyristor can be best explained by assuming it to be made up of two transistors connected back-to-back as a pair of complementary regenerative switches as shown.

A Thyristors Two Transistor Analogy

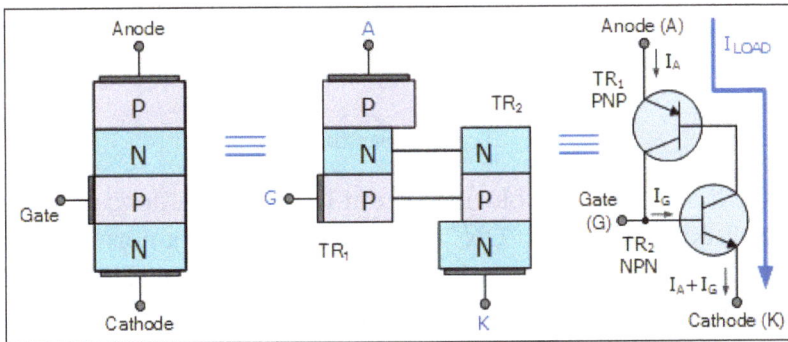

The two transistor equivalent circuit shows that the collector current of the NPN transistor TR_2 feeds directly into the base of the PNP transistor TR_1, while the collector current of TR_1 feeds into the base of TR_2. These two inter-connected transistors rely upon each other for conduction as each transistor gets its base-emitter current from the other's collector-emitter current. So until one of the transistors is given some base current nothing can happen even if an Anode-to-Cathode voltage is present.

When the thyristors Anode terminal is negative with respect to the Cathode, the centre N-P junction is forward biased, but the two outer P-N junctions are reversed biased and it behaves very much like an ordinary diode. Therefore a thyristor blocks the flow of reverse current until at some high voltage level the breakdown voltage point of the two outer junctions is exceeded and the thyristor conducts without the application of a Gate signal.

This is an important negative characteristic of the thyristor, as Thyristors can be unintentionally triggered into conduction by a reverse over-voltage as well as high temperature or a rapidly rising dv/dt voltage such as a spike.

If the Anode terminal is made positive with respect to the Cathode, the two outer P-N junctions are now forward biased but the centre N-P junction is reverse biased. Therefore forward current is also blocked. If a positive current is injected into the base of the NPN transistor TR_2, the resulting collector current flows in the base of transistor TR_1. This in turn causes a collector current to flow in the PNP transistor, TR_1 which increases the base current of TR_2 and so on.

Very rapidly the two transistors force each other to conduct to saturation as they are connected in a regenerative feedback loop that can not stop. Once triggered into conduction, the current flowing through the device between the Anode and the Cathode is limited only by the resistance of the external circuit as the forward resistance of the device when conducting can be very low at less than 1Ω so the voltage drop across it and power loss is also low.

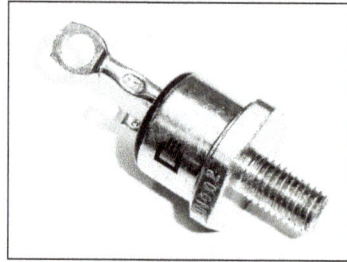

Typical Thyristor.

Then we can see that a thyristor blocks current in both directions of an AC supply in its "OFF" state and can be turned "ON" and made to act like a normal rectifying diode by the application of a positive current to the base of transistor, TR_2 which for a silicon controlled rectifier is called the "Gate" terminal.

The operating voltage-current I-V characteristics curves for the operation of a Silicon Controlled Rectifier are given as:

Thyristor I-V Characteristics Curves

Once the thyristor has been turned "ON" and is passing current in the forward direction (anode positive), the gate signal looses all control due to the regenerative latching action of the two internal transistors. The application of any gate signals or pulses after regeneration is initiated will have no effect at all because the thyristor is already conducting and fully-ON.

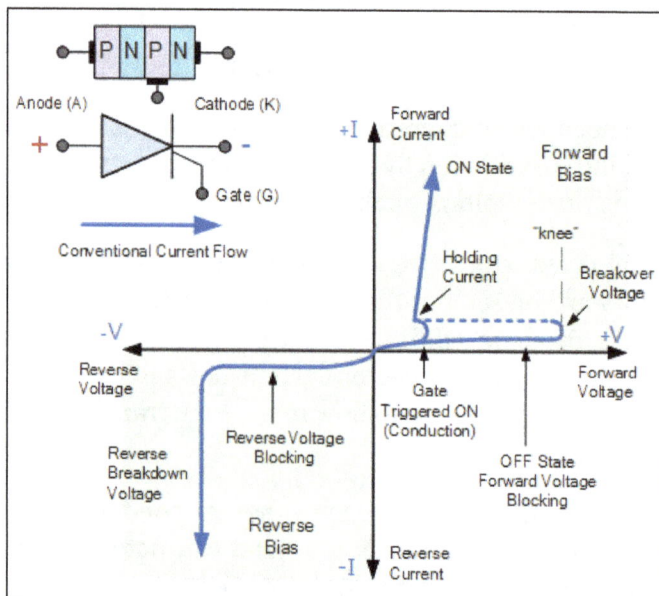

Unlike the transistor, the SCR can not be biased to stay within some active region along a load line between its blocking and saturation states. The magnitude and duration of the gate "turn-on" pulse has little effect on the operation of the device since conduction is controlled internally. Then applying a momentary gate pulse to the device is enough to cause it to conduct and will remain permanently "ON" even if the gate signal is completely removed.

Therefore the thyristor can also be thought of as a *Bistable Latch* having two stable states "OFF" or "ON". This is because with no gate signal applied, a silicon controlled rectifier blocks current in both directions of an AC waveform, and once it is triggered into conduction, the regenerative latching action means that it cannot be turned "OFF" again just by using its Gate.

Once the thyristor has self-latched into its "ON" state and passing a current, it can only be turned "OFF" again by either removing the supply voltage and therefore the Anode (I_A) current completely, or by reducing its Anode to Cathode current by some external means (the opening of a switch for example) to below a value commonly called the "minimum holding current", I_H.

The Anode current must therefore be reduced below this minimum holding level long enough for the thyristors internally latched pn-junctions to recover their blocking state before a forward voltage is again applied to the device without it automatically self-conducting. Obviously then for a thyristor to conduct in the first place, its Anode current, which is also its load current, I_L must be greater than its holding current value. That is $I_L > I_H$.

Since the thyristor has the ability to turn "OFF" whenever the Anode current is reduced below this minimum holding value, it follows then that when used on a sinusoidal AC supply the SCR will automatically turn itself "OFF" at some value near to the cross over point of each half cycle, and as we now know, will remain "OFF" until the application of the next Gate trigger pulse.

Since an AC sinusoidal voltage continually reverses in polarity from positive to negative on every half-cycle, this allows the thyristor to turn "OFF" at the 180° zero point of the positive waveform. This effect is known as "natural commutation" and is a very important characteristic of the silicon controlled rectifier.

Thyristors used in circuits fed from DC supplies, this natural commutation condition cannot occur as the DC supply voltage is continuous so some other way to turn "OFF" the thyristor must be provided at the appropriate time because once triggered it will remain conducting.

However in AC sinusoidal circuits natural commutation occurs every half cycle. Then during the positive half cycle of an AC sinusoidal waveform, the thyristor is forward biased (anode positive) and a can be triggered "ON" using a Gate signal or pulse. During the negative half cycle, the Anode becomes negative while the Cathode is positive. The thyristor is reverse biased by this voltage and cannot conduct even if a Gate signal is present.

So by applying a Gate signal at the appropriate time during the positive half of an AC waveform, the thyristor can be triggered into conduction until the end of the positive half cycle. Thus phase control (as it is called) can be used to trigger the thyristor at any point along the positive half of the AC waveform and one of the many uses of a Silicon Controlled Rectifier is in the power control of AC systems as shown.

Thyristor Phase Control

At the start of each positive half-cycle the SCR is "OFF". On the application of the gate pulse triggers the SCR into conduction and remains fully latched "ON" for the duration of the positive cycle. If the thyristor is triggered at the beginning of the half-cycle (θ = 0°), the load (a lamp) will be "ON" for the full positive cycle of the AC waveform (half-wave rectified AC) at a high average voltage of 0.318 x Vp.

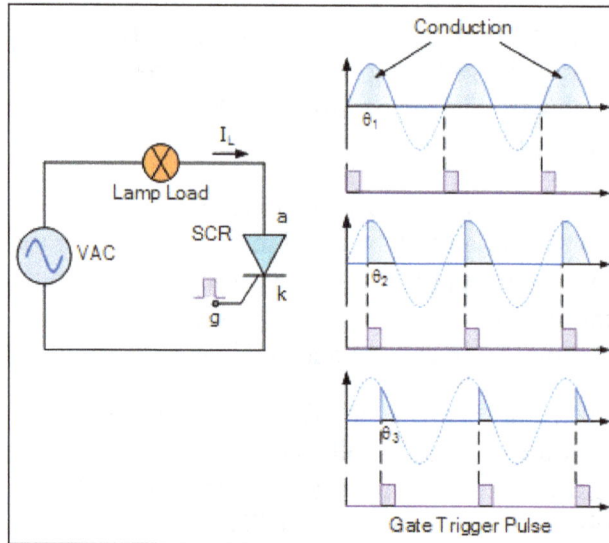

As the application of the gate trigger pulse increases along the half cycle (θ = 0° to 90°), the lamp is illuminated for less time and the average voltage delivered to the lamp will also be proportionally less reducing its brightness.

Then we can use a silicon controlled rectifier as an AC light dimmer as well as in a variety of other AC power applications such as: AC motor-speed control, temperature control systems and power regulator circuits, etc.

Thus far we have seen that a thyristor is essentially a half-wave device that conducts in only the positive half of the cycle when the Anode is positive and blocks current flow like a diode when the Anode is negative, irrespective of the Gate signal.

But there are more semiconductor devices available which come under the banner of "Thyristor" that can conduct in both directions, full-wave devices, or can be turned "OFF" by the Gate signal.

Such devices include "Gate Turn-OFF Thyristors" (GTO), "Static Induction Thyristors"

(SITH), "MOS Controlled Thyristors" (MCT), "Silicon Controlled Switch" (SCS), "Triode Thyristors" (TRIAC) and "Light Activated Thyristors" (LASCR) to name a few, with all these devices available in a variety of voltage and current ratings making them attractive for use in applications at very high power levels.

Types of Thyristors

Unidirectional Thyristor

- The thyristors which conduct in forward direction only are known as unidirectional thyristors.

- Example: SCR- Silicon Controlled Rectifier.

  LASCR-Light Activated Silicon Controlled Rectifier.

Bidirectional Thyristor

- The thyristors which can conduct in forward as well as in reverse direction are known as bidirectional thyristor.

- Example: TRIAC – TRIode AC switch.

Triggering Devices

- The devices which generate a control signal to switch the device from non-conducting to conducting state is called as triggering device.

- Example: Diode AC Switch-DIAC.

- UJT – UniJunction Transistor.

- SUS – Silicon Unilateral SwitchSBS – Silicon Bilateral Switch.

SILICON CONTROLLED RECTIFIER APPLICATIONS

Silicon Controlled Rectifier as a Switch

The SCR has only two states, namely; ON state and OFF state and no state in between. When appropriate gate current is passed, the SCR starts conducting heavily and remains in this position indefinitelyeven if the gate voltage is removed. This corresponds to the ON condition. However, when the anode current is reduced to the holding current, the SCR is turned OFF. It is clear that behaviour of SCR is similar to a mechanical switch. As SCR is an electronic device, therefore, it is more appropriate to call it an electronic switch.

SCR Switching

We have seen that SCR behaves as a switch i.e. it has only two states; ON state and OFF state.

SCR turn-on Methods

In order to turn on the SCR, the gate voltage V_G is increased upto a minimum value to initiate triggering.

This minimum value of gate voltage at which SCR is turned ON is called gate triggering voltage V_{GT}. The resulting gate current is called gate triggering current I_{GT}.

Thus to turn on an SCR all that we have to do is to apply positive gate voltage equal to V_{GT} or pass a gate current equal to I_{GT}.

For most of the SCRs, V_{GT} = 2 to 10 V and I_{GT} = 100 µA to 1500 mA.

D.C. Gate Trigger Circuit

A typical circuit used for triggering an SCR with a d.c. gate bias.

When the switch is closed, the gate receives sufficient positive voltage (=V_{GT}) to turn the SCR on.

The resistance R_1 connected in the circuit provides noise suppression and improves the turn-on time. The turn-on time primarily depends upon the magnitude of the gate current. The higher the gate-triggered current, the shorter the turn-on time.

A.C. Trigger Circuit

An SCR can also be turned on with positive cycle of a.c. gate current. Figure shows such a circuit. During the positive half-cycle of the gate current, at some point =, the device is turned on as shown in figure.

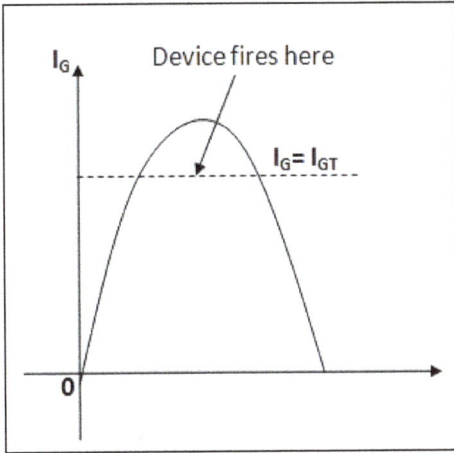

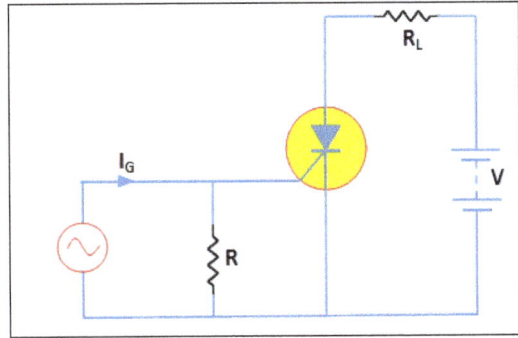

SCR Turn-off Methods

The SCR turn-off poses more problem than SCR turn-on.

It is because once the device is ON, the gate loses all control. There are many methods of SCR turn-off but only two will be discussed.

Anode Current Interruption

When the anode current is reduced below a minimum value called holding current, the SCR turns off. The simple way to turn off the SCR is to open the line switch S as shown in figure.

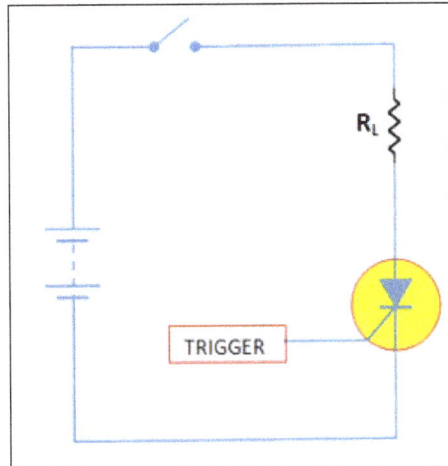

Forced Commutation

The method of discharging a capacitor in parallel with an SCR to turn off the SCR is called forced commutation.

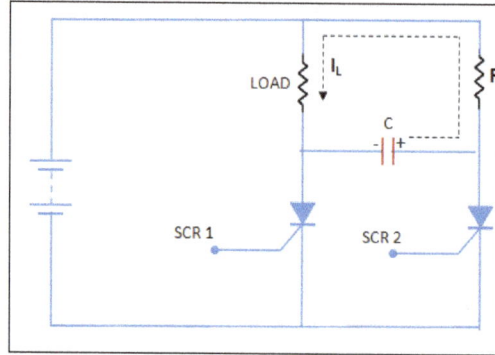

The forced commutation of SCR where capacitor C performs the commutation.

Assuming the SCRs are switches SCR1 ON and SCR2 OFF, current flows through the load and C as shown in figure.

When SCR2 is triggered ON, C is effectively paralleled across SCR1.

The charge on C is then opposite to SCR1's forward voltage, SCR1 is thus turned OFF and the current is transferred to R-SCR2 path.

Silicon Controlled Rectifier in AC Circuits

Basic Resistive Control

Thyristors are generally used in AC power control circuits such as lighting dimmers, AC motor speed controls, heaters etc. where mains (line) voltages are used for loads of many watts, or often kilowatts. The aim of AC Control is to trigger the SCR part way through each AC cycle so that the load current through the SCR is switched off for part of the AC cycle, so restricting the average current flowing through the SCR, and hence the average power delivered to the load.

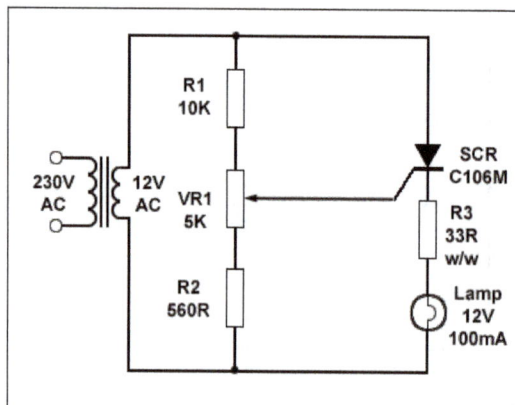

Basic Resistive Control Circuit.

The simplest way of achieving this is illustrated in figure, where the thyristor is switched on by applying a low voltage sine wave (derived from the AC input by a simple resistor network containing a variable potentiometer) to the gate terminal of the SCR. Note that because the gate input wave is derived from the AC flowing through the SCR, it will consist only of rectified half wave pulses. The effect of this input wave is that the SCR will switch on only as the gate waveform reaches the SCR firing potential, which happens part way through each positive half cycle of the AC wave. Once the thyristor is switched on it continues to conduct until the AC wave reduces to just above zero volts, when the current flowing between anode and cathode falls to a value less than the 'holding current' threshold. The thyristor then remains in a non conducting state during the negative half cycle of the AC wave as it is now reverse biased (in reverse blocking mode) during the remainder of the AC cycle. When the next positive half cycle starts the thyristor remains in a non-conducting state until the trigger waveform at the gate terminal reaches its firing potential once more.

The time or phase angle at which the SCR will be triggered can be varied by changing the amplitude of the gate waveform. As can be seen from the animation in figure the smaller the gate signal amplitude, the later the SCR switches on. Changing the amplitude of the trigger waveform therefore controls the switch-on time of the SCR. Note however that as a thyristor is basically a rectifier diode it only conducts during half of the AC cycle, a single SCR can therefore only deliver 50% of the available AC power. Also, in using this very basic form of control, the current flow through the SCR is only controllable over half of the positive half cycle, that is a quarter of the full AC cycle. It can be seen that once the switch-on time reaches the peak amplitude of the AC wave it cannot be adjusted further, as the peak amplitude of the trigger waveform will no longer reach the SCR gate firing potential and so will not trigger the SCR after this point.

Full Wave SCR Control

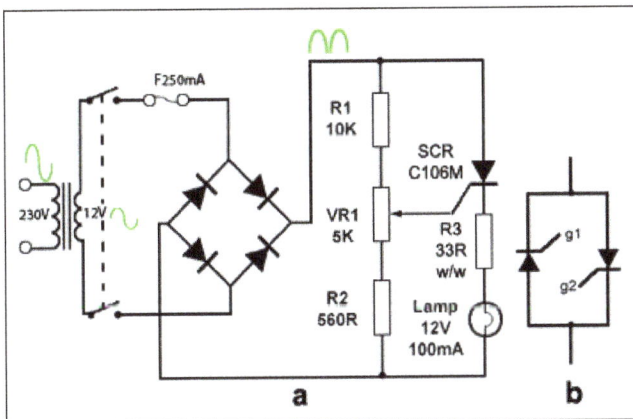

Full Wave SCR Control Methods.

The basic SCR operation can be considerably improved with some simple modifications. Perhaps the greatest drawback of the simple resistive control is that the range of adjustment could only cover 25% of the whole AC wave. This is due to the diode action of the SCR only conducting during the positive going half of the AC wave. To allow conduction during the negative going half of the AC wave, the AC can be rectified using a full wave rectifier, as shown in figure. As both halves of the AC wave will now be positive going, the range of adjustment is now improved to nearly 50%. An alternative is to use a second SCR connected in anti-parallel as shown in figure. so that one SCR conducts during positive half cycles, and the other SCR during negative half cycles. However this parallel arrangement of SCRs can also be obtained simply by using a single Triac instead of two SCRs.

SCR Phase Control

To achieve control over virtually 100% of the AC wave, phase control simply replaces one of the resistors in the resistive control circuit with a capacitor. This now converts the resistor network into a variable low pass filter that will shift the phase of the AC wave applied to the gate. basically, the values of C and R are chosen so that adjustment of R1 will provide a phase shift from 0° to nearly 90°. To be effective, the variation of R1 needs to give sufficient change in the behaviour of the load device (in this case a 12volt 100mA lamp). As well as shifting the phase of the gate waveform however, the RC filter will also be altering the amplitude of the gate waveform, so the amplitude of the gate waveform also needs to be kept above the firing potential of the SCR type chosen, for switching to take place. From these conditions it can be seen that calculation of suitable values for R and C to provide appropriate control, depend on both phase and amplitude so can get quite complex. Therefore some practical experimentation with R and C values is also most likely to be necessary.

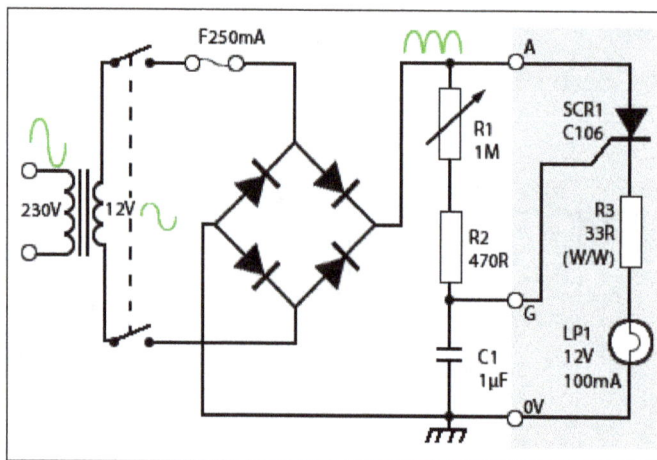

SCR Phase Control Demo Circuit.

SCR Level Control

Another way of switching on the SCR at the appropriate part of the AC cycle is to apply a DC voltage to the gate during the time the SCR is required to conduct. The DC applied to the gate will therefore be a variable width pulse having a voltage level sufficient to cause the SCR to conduct. These pulses must be synchronised with the rectified AC wave so that they always start and end at the correct time relative to the AC waveform.

The animation in figure illustrates the basic method of triggering an SCR using level control. The SCR is triggered (switched on) for a period during each rectified AC half cycle by a voltage Vg applied to the SCR gate. The SCR turns off at the end of each half cycle as the voltage across the SCR falls to near zero, which also coincides with the end of the trigger pulse Vg. The DC pulses may be generated digitally, using a computer output or by using a discrete component circuit such as that shown below in figire, which uses a 555 timer based monostable. This circuit offers a simple and inexpensive method of demonstrating SCR operation using only low voltages. Two power supplies are used, the shaded area of figure is the AC demonstration power supply, which isolates the demonstration circuit from the mains (line) supply. The control section of the circuit must be supplied with a DC voltage of between 5V and 12V. This can be from either a separate DC power supply (e.g. a 'Wall Wart'), a dedicated IC regulated supply, or a battery. The control section of the circuit (black) is also isolated from the AC section (red) by two optocouplers, IC1 and IC3. Because this circuit is already isolated from mains voltage by T1, it would seem unnecessary to use a second method of isolation in IC1, However the main function of IC1 is not isolation in this case, but to act as a zero crossing detector.

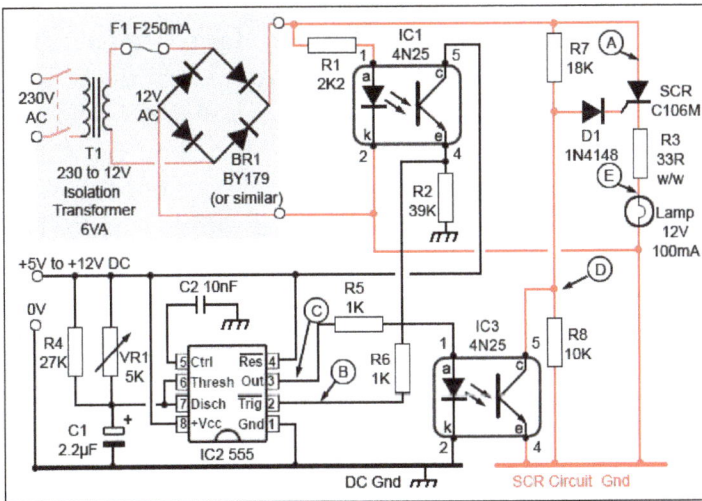

SCR Level Triggering Circuit.

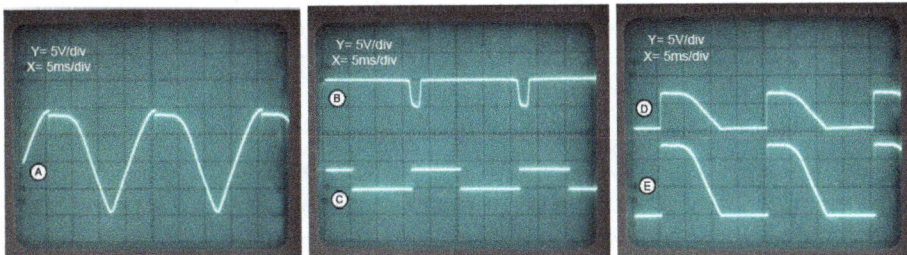

SCR Level Triggering Waveforms.

Level Triggering Demonstration Circuit

The circuit in figure switches on the SCR at a time chosen by the setting of VR1 during each positive AC half cycle from the low voltage power supply (waveform A). The SCR switches off again as the rectified AC voltage reduces to near zero at the end of each half cycle. The control circuit is based around a 555 timer IC operating in monostable mode, and two 4N25 opto couplers.

As well as isolating the 555 circuitry from the incoming AC, IC1 (4N25) provides a synchronising pulse (waveform B in figure). This is achieved by biasing IC1 in common collector mode so that its output transistor conducts for most of the full wave AC input, producing a high (5V) voltage at pin 4, but turns off as the AC wave approaches 0V, producing a 0V output at pin 4 of IC1. These pulses are used to trigger the 555 monostable (IC2) at the start of each half cycle.

Each time IC2 is triggered its output on pin 3 goes high for a time set by the time constant created by variable resistor VR1 and the timing capacitor C1. Notice that VR1 is also connected in parallel with a 27K resistor R4. The purpose of this is to achieve a more accurate time constant than is possible using only the preferred values of VR1 and C1. It would also be possible to fit a preset resistor in place of R4 to obtain the exact duration for the high level trigger pulse produced by IC2.

Notice that the trigger pulse produced by IC2 (waveform C in figure) goes high immediately a synchronising pulse is received, which would turn the SCR on at the start of the half cycle. Also when the trigger pulse returns low this would not switch off the SCR, it would continue to conduct until the end of the half cycle; this is not what is needed. However, waveform C is inverted by the action of the optocoupler IC3, because its output transistor is connected in common emitter mode. Therefore the SCR is triggered during the latter period of the rectified AC half cycle, (waveform D in figure). Notice that waveform D does not look like the inverse of waveform C because, as soon as the SCR is triggered the gate input (together with the anode and cathode) follows the shape of the rectified AC wave from the moment of triggering to the time it reaches 0V.

SCR Pulse Triggering

Using level triggering has the drawback of creating gate current throughout the 'on' period of the SCR. This creates unnecessary gate current and in high power application can add to heat generated at junction 2 of the SCR, which in turn may reduce long term reliability.

A modification to the circuit shown in figure is illustrated in figure. This circuit generates a single narrow pulse (about 4µs in duration) to trigger the SCR at the chosen firing angle, the SCR then continues to conduct until the forward current falls to less than the holding current value at around 0V so greatly reducing the average gate current.

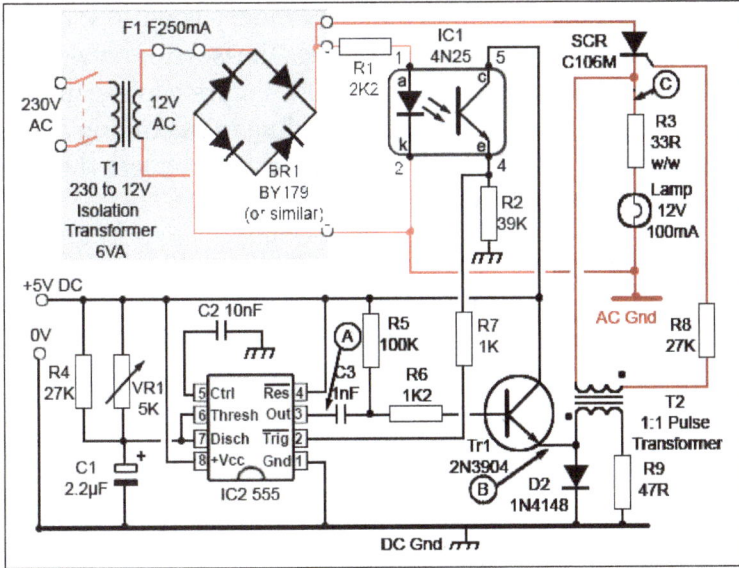

SCR Pulse Triggering Circuit.

Pulse Triggering Circuit Works

The portion of figure shown in pale grey works in the same way as already described for figure; the output of IC2 (the monostable) consists of variable width positive pulses (waveform A shown in figure) where the falling edge of each pulse defines the firing angle of the SCR. (Note that in the level triggering circuit this waveform is inverted before being applied to the gate, so that the falling edge becomes a rising edge to trigger the SCR). In figure before the output of IC2 is inverted, it is differentiated by C3 and R5 to produce a series of narrow 4µs positive and negative going pulses corresponding to the rising and falling edges of waveform A. These narrow pulses are fed to the common collector (emitter follower) driver transistor Tr1 via R6. Diode D2 at Tr1 emitter removes the positive going pulses (apart from a small residue due to the forward junction potential of the diode).

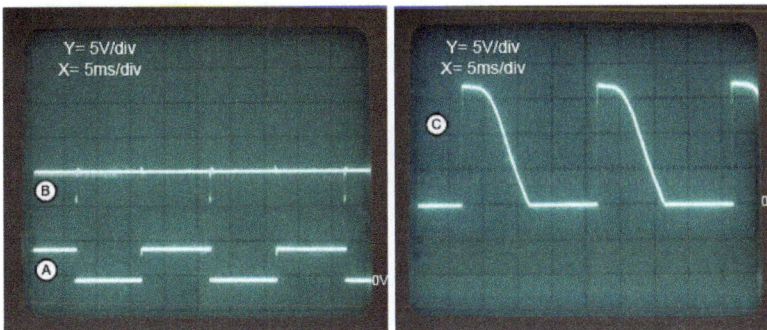

SCR Pulse Triggering Waveforms.

The negative going pulses (waveform B) at Tr1 emitter are inverted by the 1:1 pulse transformer T2 by connecting T2 secondary in anti-phase to the T2 primary (notice the phase indicator dots next to the primary and secondary windings) so producing positive going trigger pulses for the SCR. T2 also acts as the isolator between the low voltage DC control circuit and the higher voltage AC SCR. Figure waveform C shows the SCR cathode waveform, the fast rising edge corresponding to the timing of the trigger pulse delivered to the gate via R8 current limiting resistor; this reduces the current delivered by each trigger pulse to around 100µA.

Both the level triggering and pulse triggering circuits provide reliable triggering and adjustment over nearly the whole 360° of the 50Hz AC wave. Some adjustment of the monostable time constant may be necessary for 60Hz operation. The DC supply voltage level is not critical, between about 5V and 12V.

Synchronous (Zero Crossing) Switching

The AC output waveform when the SCR is switched on during each positive half cycle of the AC wave, has a very fast rise time, as the current through the SCR suddenly switches from zero to the instantaneous value of the AC wave. When used with a 230V AC supply this sudden change can be around 325V (the peak value of the AC wave). The waveform may also be a sharp triangular spike if the SCR switches on after the peak value of the wave has occurred. In any case the AC voltage waveform produced by the SCR action will be rich in harmonics, that can generate a serious level of electromagnetic interference (e.m.i.) causing problems not only to other connected circuitry; the interference can also radiate to other nearby electronics as radio frequency interference (r.f.i.) as the harmonics produced can extend well into the radio frequency bands. To avoid these problems alternative control methods may be used. One such method, called 'Synchronous or Zero Crossing Switching' is to only allow thyristors to switch when the mains waveform is at, or very close to zero volts. The thyristor is then switched on for a number of cycles and then switched off again (as the AC voltage passes through 0V) for another number of cycles. The ratio of on to off cycles can then be altered to provide a variation of average power supplied to the load.

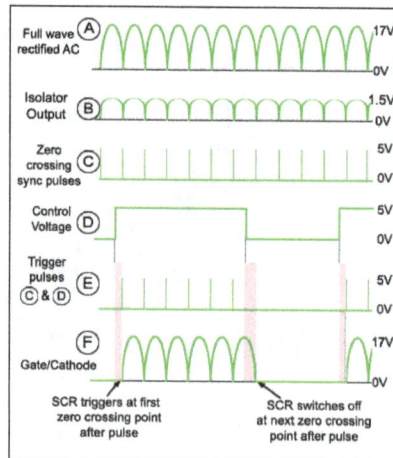

SCR Zero Crossing Waveforms.

Waveform A in figure shows the 18Vpp 100Hz waveform applied to the zero crossing circuit from the full wave rectified AC power supply and bridge rectifier.

Waveform B is a series of 5V pulses, derived from IC1 optocoupler. As the optocoupler transistor is turned on for most of the positive half cycle of the AC input, this makes the emitter high apart from a narrow pulse as the emitter falls from 5V to 0V each time the AC input falls to 0V. These pulses are therefore synchronised to the zero voltage point of waveform A.

However, as positive going trigger pulses are needed to trigger the SCR, the pulses at B are inverted by Tr1 to create waveform C.

Waveform D is the output of a free running 555 astable oscillator IC2, which produces square pulses at a pulse repetition frequency of about 7Hz and a variable duty cycle adjustable by VR1. This waveform is used to control the ratio of the on and off times of the SCR. As the SCR will be high (on) for a number of 100Hz half cycles, then low (off) for a number of half cycles. The mark to space ratio of the square wave produced by IC2 is adjustable by VR1 to produce an on time of between about 20% and 90% of the periodic time of the astable output.

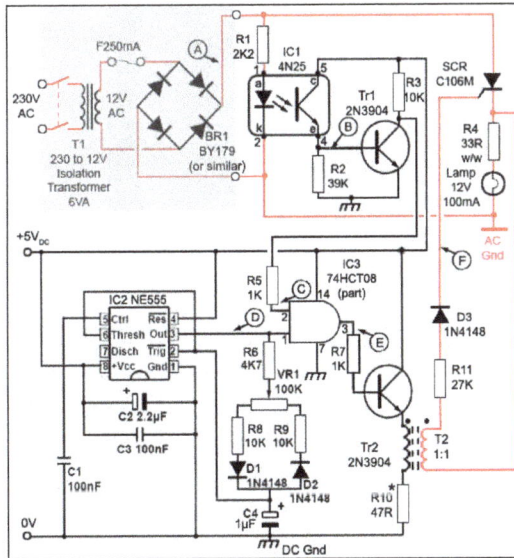

SCR Zero Crossing Control Circuit.

The outputs of Tr1 (waveform C) and IC2 (waveform D) are applied to the two inputs of the AND gate (IC3). The output of IC3 goes to logic 1 only when both inputs are at logic 1. This produces a series of narrow positive going trigger pulses (waveform E) to trigger the SCR only at the start of those half cycles whilst waveform D is high. The trigger pulses produced are applied to T2, a 1:1 isolating pulse transformer via the emitter follower driver transistor Tr2. The secondary winding of T2 applies the trigger pulses to the gate of the SCR via a current limiter resistor R11 and diode D3. The gate waveform (waveform F) is practically identical to the output waveform at the SCR cathode as there is only a small voltage difference between the gate and cathode of the SCR.

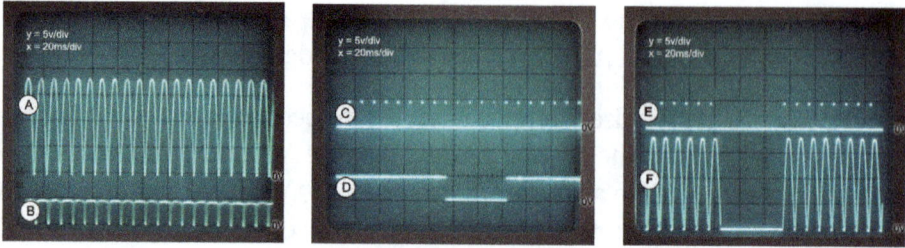

Waveforms.

SCR Zero Crossing Circuit Operation

This demonstration circuit again uses the low voltage (12VRMS) full wave rectified AC supply.

SCR Zero Crossing Breadboard Circuit.

Figure uses two different methods of isolation and demonstrates how the zero crossing control method may be achieved using standard components. It is not meant to represent any particular commercially available solution, nor is it meant to represent the best available method. The purpose of the SCR gate drive circuits discussed is to provide useful demonstrations of commonly used drive techniques and a low voltage environment for relevant experimentation. They can be inexpensively built on standard breadboard or strip board as shown in figure to serve as useful demonstrations.

Silicon Controlled Rectifier in DC Circuits

DC Power Switching

Thyristors can be used to control either AC or DC loads and can be used to switch low voltage low current devices as well as very large currents at mains (line) voltages. A simple example of a thyristor controlling a DC load, such as a small DC motor is illustrated in figure. The motor here is connected to a 12V DC supply via a BT151 thyristor, but will not run until the thyristor is made to conduct. This is achieved by momentarily closing the 'start' switch, which provides a pulse of current to the gate terminal of the thyristor.

The motor now runs as the thyristor switches on and its resistance is now very low.

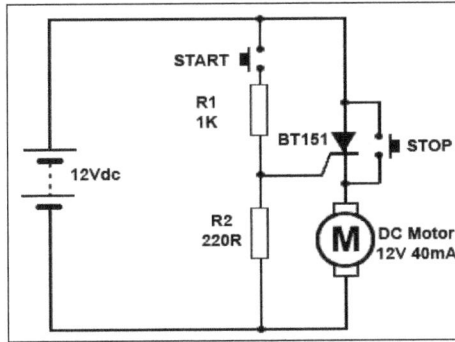

When the start switch returns to its normally open state, there is no longer any gate current but the thyristor continues to conduct, and in a DC circuit, current will continue to flow and the motor continues to run. Any further operations of the start switch now have no effect. The thyristor will only switch off if current flow reduces to a value below the thyristor's holding current threshold.

This is achieved by shorting out the thyristor by momentarily closing the 'stop' switch. The circuit current now flows through the stop switch rather than through the thyristor, which instantly turns off, as the SCR current is now reduced to less than the holding current value. Stopping the motor could also be achieved by using a normally closed switch in series with the thyristor, which when pressed, would also temporarily prevent current flow through the thyristor long enough for the thyristor to turn off.

Although this simple circuit works, as can be seen in the video accompanying figure it is not difficult to imagine simpler ways of switching a small motor on and off. However the principle is useful in situations such as using a computer to control a DC motor. The small current produced by the computer's output is used to trigger a thyristor (usually via an opto coupled device to provide electrical isolation). The thyristor can then supply the motor or other device with whatever higher value of current is required. The use of a thyristor could, with some appropriate extra circuitry, also allow for remote switching of a circuit or device, triggered for example by a radio signal.

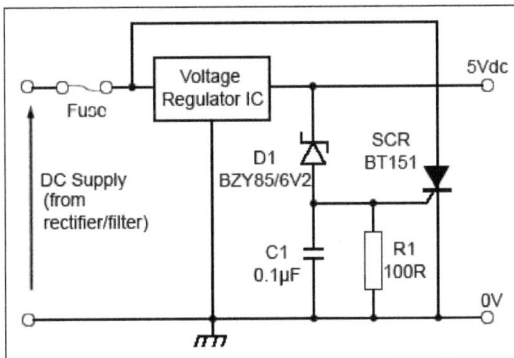

Crowbar Over Voltage Protection.

SCR Crowbar Circuits

Another DC operation using thyristors is the 'Crowbar' circuit, used as an over voltage protection device. The circuit is called a crowbar as its action is about as subtle as a swift blow with a crowbar. Such circuits may often be found preventing power supply circuits from outputting a higher than normal voltage under fault conditions.

The basic idea is that if, for example a fault in a DC power supply line causes the output to rise above its specified voltage value, this 'over voltage' is sensed and causes a normally non conducting SCR connected between the power supply output and ground to switch on very rapidly. This can have different protective actions, the simplest of which, as illustrated in figure is to blow a fuse and so switch off the power completely, requiring the attention of a service technician to get the circuit working again. This is often chosen as the safest option as the cause of the original over voltage should be examined and eliminated before the circuit is allowed to work again.

The output of a regulated 5V DC supply is sensed by D1, a 6.2V Zener diode, the anode of which is held at a voltage close to 0V by R1. This 100Ω resistor ensures that if the 5V supply line rises above its specified limit, sufficient current flows through the Zener diode to provide enough current at the SCR gate to switch on the SCR. Care must also be taken to ensure that the SCR is not triggered accidentally by any fast voltage spikes appearing on the 5V line, due for example to other switching devices in the circuit being supplied. C1 is therefore connected between the gate and cathode of the SCR to reduce the amplitude of any very short interference pulses, provided they do not exist long enough to charge C1 to a high enough level to trigger the SCR.

The reason for using a thyristor to blow the fuse is that fuses do not blow immediately, they operate by blowing when excessive current flows for long enough so that the fusible element heats up and melts. This may take long enough for the excessive voltage to have already destroyed a number of semiconductor components. The thyristor however has a very fast switch on time (about 2μs for the BT151) so that during the short time between the over voltage occurring and the fuse blowing, the entire power supply output current will be flowing through the thyristor, rather than through the circuit being supplied.

Silicon Controlled Rectifier as Half Wave Rectifier

SCRs are very useful in ac circuits where they may serve as rectifiers whose output current can be controlled by controlling the gate current. An example of this type of application is the use of SCRs to operate and control dc motors or dc load from an ac supply. The circuit using an SCR as an half-wave rectifier is shown in figure. The ac supply to be rectified is applied to the primary of the transformer ensuring that the negative voltage appearing at the secondary of the transformer is less than reverse breakdown voltage of the SCR. The load resistance R_L is connected in series with anode. A variable resistance r is inserted in the gate circuit for control of gate current.

If the angle at which the SCR starts conducting (i.e. firing angle) is a, the conduction will take place for $(\Pi - \alpha)$ radians.

The average output from such a half-wave rectifier connected to a secondary voltage of:

$v = V_{max} \sin \theta$ *is given by an expression*

Average output voltage,

$$V_{av} = V_{MAX}/2\Pi(1 + \cos \alpha)$$

Average current,

$$I_{av} = V_{MAX}/2\Pi R_L (1 + \cos \alpha)$$

Thus the desired value of average current, I_{av} can be obtained by varying firing angle α.

$$I_{av} = V_{MAX}/\Pi R_L \text{ when } \alpha = 0$$

$$I_{av} = V_{MAX}/2\Pi R_L \text{ when } \alpha = \Pi/2$$

That is average current decreases with the increase in value of firing angle α.

The worth noting point is that in an ordinary half-wave rectifier using a P-N diode, conduction current flows during the whole of the positive cycle whereas in SCR half-wave rectifier the current can be made to flow during the part or full of the positive half cycle by adjustment of gate current. Hence SCR operates as a controlled rectifier and hence the name silicon controlled rectifier.

The output voltage from the SCR rectifier is not a purely dc voltage but also consists of some ac components, called the ripples, along it. The ripple components are undesirable and need to be removed or filtered out. This is accomplished by placing a filter circuit between the rectifier and load.

During the negative half cycles of ac voltage appearing across the secondary, the SCR does not conduct regardless of the gate voltage, because anode is negative with respect to cathode and also peak inverse voltage is less than the reverse breakdown voltage. The SCR will conduct during the positive half cycles provided appropriate gate current is made to flow. The gate current can be varied with the help of variable resistance r inserted in the gate circuit for this purpose. The greater the gate current, the lesser will be the supply voltage at which SCR will start conducting.

Assume that gate current is such that SCR starts conducting at a positive voltage V, being less than peak value of ac voltage, V_{max}. From figure, it is clear that SCR will start conducting, as soon as the secondary ac voltage becomes V in the positive half cycle, and will continue conducting till ac voltage becomes zero when it will turn-off. Again in next positive half cycle, SCR will start conducting when ac secondary voltage becomes V volts.

Silicon Controlled Rectifier as Full Wave Rectifier

For full-wave rectification two SCRs are connected across the centre taped secondary, as shown in figure-a. The gates of both SCRs are supplied from two gate control supply circuits. One SCR conducts during the positive half cycle and the other during the negative half cycle and thus unidirectional current flows in the load circuit. The main advantage of this circuit over ordinary full-wave rectifier circuit is that the output voltage can be controlled by adjusting the gate current.

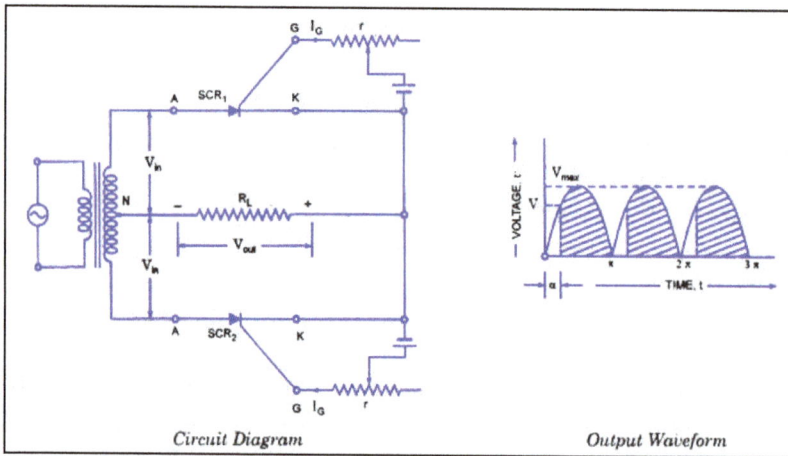

Circuit Diagram   *Output Waveform*

SCR Full-Wave Rectifier.

Now if the supply voltage v = V_{MAX} sin θ and the firing angle is a, then average voltage output will be given by the expression:

$$V_{av} = V_{MAX}/\Pi(1 + Cos\ \alpha)$$

That is, average voltage output of full-wave rectifier circuit is double of that of half-waveÂ rectifier circuit, which is obvious that:

$$I_{av} = V_{MAX}/\Pi R_L (1 + Cos\ \alpha).$$

DIAC

The DIAC is a diode that conducts electrical current only after its breakover voltage, V_{BO}, has been reached momentarily. The term is an acronym of "diode for alternating current".

When breakdown occurs, the diode enters a region of negative dynamic resistance, leading to a decrease in the voltage drop across the diode and, usually, a sharp increase in current through the diode. The diode remains in conduction until the current through it drops below a value characteristic for the device, called the *holding current*, I_H. Below this value, the diode switches back to its high-resistance, non-conducting state. This behavior is bidirectional, meaning typically the same for both directions of current.

Most DIACs have a three-layer structure with breakover voltage of approximately 30 V. Their behavior is similar to that of a neon lamp, but it can be more precisely controlled and takes place at a lower voltage.

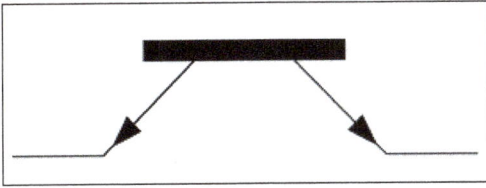

Three-layer DIAC.

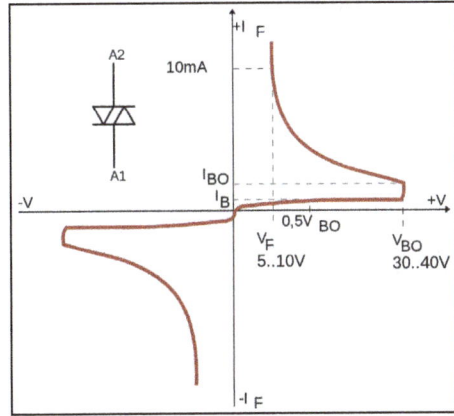

Typical DIAC voltage and current relationships. V_{BO} is the breakover voltage.

DIACs have no gate electrode, unlike some other thyristors that they are commonly used to trigger, such as TRIACs. Some TRIACs, like Quadrac, contain a built-in DIAC in series with the TRIAC's gate terminal for this purpose.

DIACs are also called "symmetrical trigger diodes" due to the symmetry of their characteristic curve. Because DIACs are bidirectional devices, their terminals are not labeled as anode and cathode but as A1 and A2 or main terminal MT1 and MT2.

SIDAC

A Silicon Diode for Alternating Current (SIDAC) is a less commonly used device, electrically similar to the DIAC, but having, in general, a higher breakover voltage and greater current handling capacity.

The SIDAC is another member of the thyristor family. Also referred to as a SYDAC (Silicon thYristor for Alternating Current), bi-directional thyristor breakover diode, or more simply a bi-directional thyristor diode, it is technically specified as a bilateral voltage triggered switch. Its operation is similar to that of the DIAC, but SIDAC is always a five-layer device with low-voltage drop in latched conducting state, more like a voltage triggered TRIAC without a gate. In general, SIDACs have higher breakover voltages and current handling capacities than DIACs, so they can be directly used for switching and not just for triggering of another switching device.

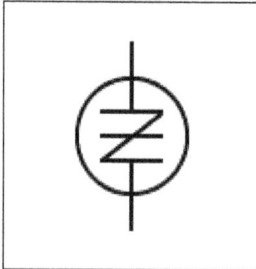

SIDAC.

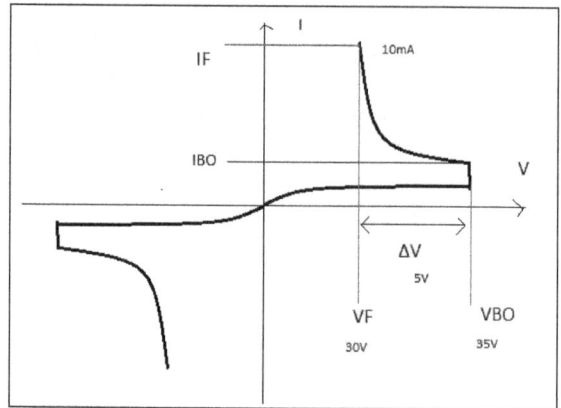

Idealized breakover diode voltage and current relationships.

The operation of the SIDAC is functionally similar to that of a spark gap, but is unable to reach its higher temperature ratings. The SIDAC remains nonconducting until the applied voltage meets or exceeds its rated breakover voltage. Once entering this conductive state going through the negative dynamic resistance region, the SIDAC continues to conduct, regardless of voltage, until the applied current falls below its rated holding current. At this point, the SIDAC returns to its initial nonconductive state to begin the cycle once again.

Somewhat uncommon in most electronics, the SIDAC is relegated to the status of a special purpose device. However, where part-counts are to be kept low, simple relaxation oscillators are needed, and when the voltages are too low for practical operation of a spark gap, the SIDAC is an indispensable component.

Similar devices, though usually not functionally interchangeable with SIDACs, are the Thyristor Surge Protection Devices (TSPD) sold under trademarks like Trisil by ST-Microelectronics and SIDACtor and its predecessor Surgector by Littelfuse. These are designed to tolerate large surge currents for the suppression of overvoltage transients. In many applications this function is now served by metal oxide varistors (MOVs), particularly for trapping voltage transients on the power mains.

TRIAC

Triacs are widely used in AC power control applications. They are able to switch high voltages and high levels of current, and over both parts of an AC waveform. This makes triac circuits ideal for use in a variety of applications where power switching is needed.

One particular use of triac circuits is in light dimmers for domestic lighting, and they are also used in many other power control situations including motor control.

As a result of their performance, trials tend to be used for low to medium power applications, leaving thyristors to be used for the very heat duty AC power switching applications.

A medium current triac.

Triac Basics

The triac is a development of the thyristor. While the thyristor can only control current over one half of the cycle, the triac controls it over two halves of an AC waveform.

As such the triac can be considered as a pair of parallel but opposite thyristors with the two gates connected together and the anode of one device connected to the cathode of the other, etc.

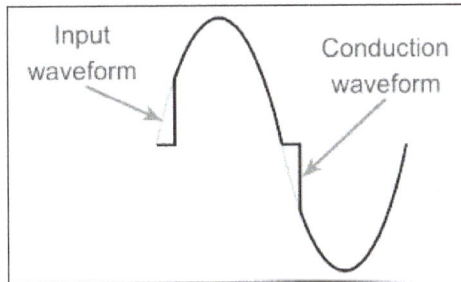

Triac switching waveform.

The fact that the triac switching action occurs on both halves of an AC waveform means that for AC power applications, the complete cycle can be used. For basic thyristor circuits, only half the waveform is used and this means that basic circuits using thyristors will not utilise both halves of the cycle. Two devices are required to utilise both halves. However the triac only requires one device to control both halves of the AC waveform.

Triac Symbol

The basic triac symbol used on circuit diagram indicates its bi-directional properties. The triac symbol can be seen to be a couple of thyristor symbols in opposite senses merged together.

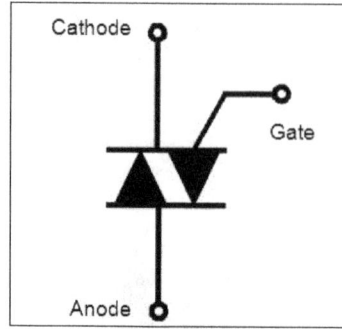
Triac circuit symbol.

Like a thyristor, a triac has three terminals. However the names of these are a little more difficult to assign, because the main current carrying terminals are connected to what is effectively a cathode of one thyristor, and the anode of another within the overall device. There is a gate which acts as a trigger to turn the device on. In addition to this the other terminals are both called Anodes, or Main Terminals These are usually designated Anode 1 and Anode 2 or Main Terminal 1 and Main Terminal 2 (MT1 and MT2). When using triacs it is both MT1 and MT2 have very similar properties.

Working of TRIAC

Before looking at how a triac works, it helps to have an understanding of how a thyristor works. In this way the basic concepts can be grasped for the simpler device and then applied to a triac which is more complicated.

For the operation of the triac, it can be imagined from the circuit symbol that the triac consists of two thyristors in parallel but around different ways. The operation of the triac can be looked on in this fashion, although the actual operation at the semiconductor level is rather more complicated.

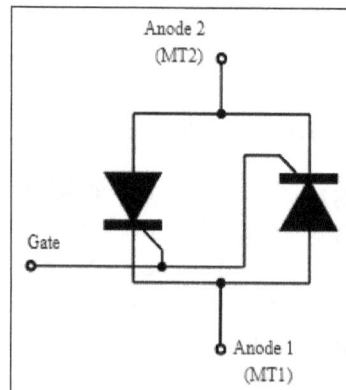
Triac equivalent circuit.

The triac structure is shown below and it can be seen that there are several areas of n-type and p-type material that form what is effectively a pair of back to back thyristors.

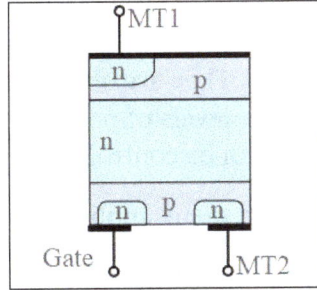

Triac basic structure.

The triac is able to conduct in a number of ways - more than the thyristor. It can conduct current irrespective of the voltage polarity of terminals MT1 and MT2. It can also be triggered by either positive or negative gate currents, irrespective of the polarity of the MT2 current. This means that there are four triggering modes or quadrants:

- *I+ Mode* MT2 current is +ve, gate current is +ve.

- *I- Mode* MT2 current is +ve, gate current is -ve.

- *III+ Mode:* MT2 current is -ve, gate current is +ve.

- *III- Mode:* MT2 current is -ve, gate current is −ve.

It is found that the triac trigger current sensitivity is greatest when the MT2 and gate currents are both of the same polarity, i.e. both positive or both negative. If the gate and MT2 currents are of the opposite polarity then the sensitivity is typically about half the value of when they are the same.

The typical IV characteristic of a triac can be seen in the diagram below with the four different quadrants labelled.

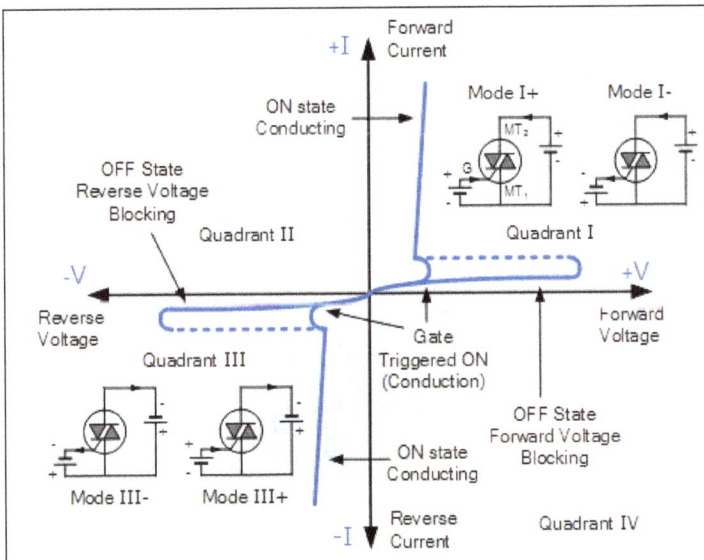

Triac IV characteristic.

Triac Applications

Triacs are used in many applications. They are often used in low to medium power AC switching requirements. Where large levels of power need to be switched, two thyristors / SCRs tend to be used as they can be controlled more easily.

Nevertheless triacs are widely used in many applications:

- Lighting control - especially domestic dimmmers.

- Control of fans and small motors.

- General AC switching and control.

There are naturally many other triac applications, but these are some of the most common.

In one specific application, triacs can be included in modules called solid state relays. Here an optical version of this semiconductor device is activated by an LED light source turning the solid state relay on according to the input signal.

Typically within solid state relays, the LED light or infrared source and the optical triac are contained within the same package, sufficient isolation being provided to withstand high voltages which may extend to hundreds of volts or possibly even more.

Solid state relays come in many forms, but those used for AC switching may use a triac.

Using Triacs

There are a number of points to note when using triacs. Although these devices operate very well, to get the best performance out of them it is necessary to understand a few hints on tips on using triacs.

It is found that because of their internal construction and the slight differences between the two halves, triacs do not fire symmetrically. This results in harmonics being generated: the less symmetrical the triac fires, the greater the level of harmonics that are produced. It is not normally desirable to have high levels of harmonics in a power system and as a result triacs are not favoured for high power systems. Instead for these systems two thyristors may be used as it is easier to control their firing.

Internal circuitry of triac light dimmer.

To help in overcoming the problem of the triac non-symmetrical firing, and the resulting harmonics, a device known as a diac (diode AC switch) is often placed in series with the gate of the triac. The inclusion of this device helps make the switching more even for both halves of the cycle. This results from the fact that the diac switching characteristic is far more even than that of the triac. Since the diac prevents any gate current flowing until the trigger voltage has reached a certain voltage in either direction, this makes the firing point of the triac more even in both directions.

Triac Circuit Examples

There are many ways in which triacs can be used. The two examples below give a taste of what can be done with these semiconductor devices.

- Simple triac switch circuit: The triac can function as a switch - it could enable a trigger pulse of a low power switch to turn the triac on to control a much higher power levels that might be possible with a simple switch.

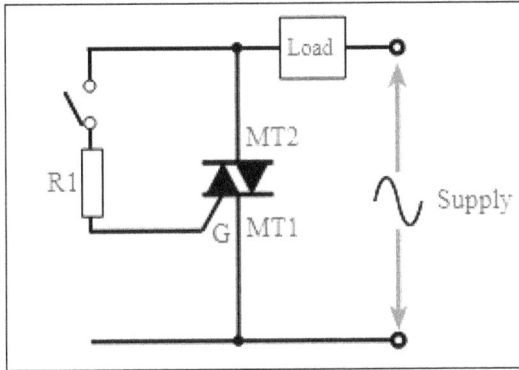

Simple triac switch circuit.

- Triac variable power or dimmer circuit: One of the most popular triac circuits varies the phase on the input of the triac to control the power that can be dissipated into load.

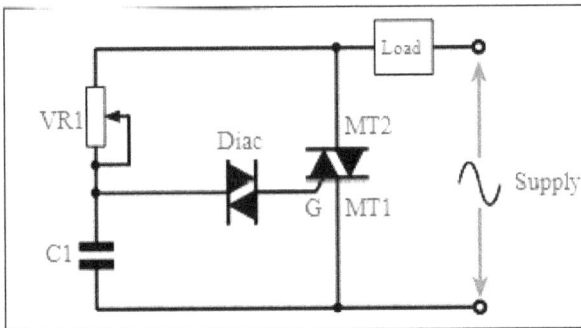

A basic triac circuit using phase of input waveform to control dissipated power in the load.

There are many more triac circuits that can be used. The device is very versatile and can be used in a variety of circuits, typically to provide various forms of AC switching.

Triac Specifications

Triacs have many specifications that are very similar to those of thyristors, although obviously they are intended for triac operation on both halves of a cycle and need to be interpreted as such.

However as their operation is very similar, so too are the basic specification types. Parameters like the gate triggering current, repetitive peak off-state voltage and the like are all required when designing a triac circuit, ensuring there is sufficient margin for the circuit to operate reliably.

Triacs are ideal devices for use in many AC small power applications. Triac circuits for use as dimmers are widespread and they are simple and easy to implement. When using triacs, diacs are often included in the circuit as mentioned above to help reduce the level of harmonics produced.

References

- Thyristor, power: electronics-tutorials.ws, Retrieved 13 July, 2019
- Thyristor: completepowerelectronics.com, Retrieved 14 August, 2019
- Scr-as-a-switch: electronicspost.com, Retrieved 15 January, 2019
- Thyristors-62, semiconductors: learnabout-electronics.org, Retrieved 16 February, 2019
- Thyristors-61, semiconductors: learnabout-electronics.org, Retrieved 17 March, 2019
- Scr-as-half-wave-rectifier: circuitstoday.com, Retrieved 18 April, 2019
- Full-wave-rectifier-using-scr: circuitstoday.com, Retrieved 19 May, 2019
- What-is-a-triac, electronic-components: electronics-notes.com, Retrieved 20 June, 2019

5

Rectifiers

The electrical devices which convert alternating current to direct current are known as rectifiers. Common types of rectifiers include single phase uncontrolled rectifiers, single phase controlled rectifiers and three phase rectifiers. The chapter closely examines these types of rectifiers to provide an extensive understanding of the subject.

RECTIFICATION

Alternating current is suitable for both heating and lighting because the heating effect of a current is independent of its direction. However, in many practical systems, the use of alternating current will not be possible. For example, in large motors and electric railway systems, we will need the operating current to flow in one direction consistently and alternating current is not able to fulfil that criterion.

But direct currents are harder to generate than alternating currents and alternaing voltages are more convenient to step up and to step down, the process of converting a.c. to d.c. by a rectifier is important.

Rectification is the process in which an alternating current is forced to only flow in one direction.

Half-wave Rectification

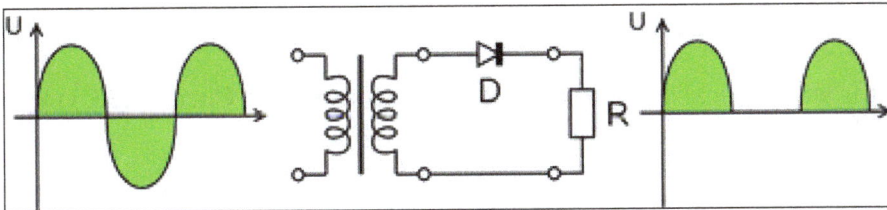

From the circuit shown in the image, a diode is connected to the circuit. Since the diode only allows current to flow in ONE direction, the current in the other direction will be zero (recall that the direction of current flow in A.C. reverses periodically). This means that half of the power is lost in half-wave rectification.

Full-wave Rectification

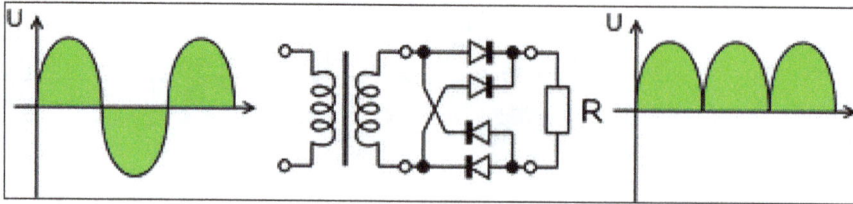

From the circuit shown in the image, 4 diodes are connected to the circuit. It might seem confusing at first but it is incredibly simple. Walkthrough of the current flow in full-wave rectification circuit:

- Since the diodes are arranged in a vertical straight line, we shall label them as 1, 2, 3, 4 with 1 being the top diode and 4 being the bottom diode. Let's start with the top part of the circuit where the current path immediately branches out to two diodes with opposite direction (diode 1 and 3).

- We'll start with the current flowing towards the right. Diode 1 will allow the current to flow while diode 3 will not.

- After passing through diode 1, the current will encounter diode 2 and the resistor (load). Since diode 2 will block the current from flowing through, the current will then pass through the resistor.

- After passing through the resistor, the current reaches another crossroad (diode 3 and diode 4). The current will not flow through diode 3 even though the direction is correct. This is because current does not flow from low potential to high potential. (After passing through the resistor, the current will have a potential loss).

- Hence, the current will flow through diode 4 and complete the circuit.

- You can repeat this for the other direction of A.C. flow.

RECTIFIERS

A rectifier is an electrical device that converts an Alternating Current (AC) into a Direct Current (DC) by using one or more P-N junction diodes.

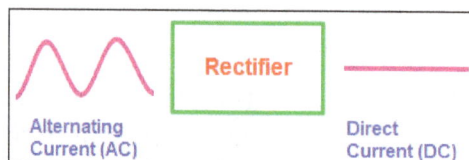

When the voltage is applied to the P-N junction diode in such a way that the positive terminal of the battery is connected to the p-type semiconductor and the negative terminal of the battery is connected to the n-type semiconductor, the diode is said to be forward biased.

When this forward bias voltage is applied to the P-N junction diode, a large number of free electrons (majority carriers) in the n-type semiconductor experience a repulsive force from the negative terminal of the battery similarly a large number of holes (majority carriers) in the p-type semiconductor experience a repulsive force from the positive terminal of the battery.

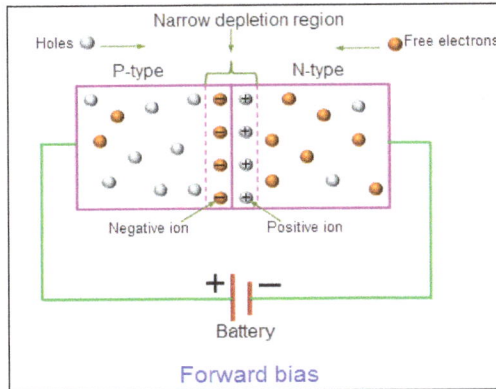

As a result, the free electrons in the n-type semiconductor start moving from n-side to p-side similarly the holes in the p-type semiconductor start moving from p-side to n-side.

We know that electric current means the flow of charge carriers (free electrons and holes). Therefore, the flow of electrons from n-side to p-side and the flow of holes from p-side to n-side conduct electric current. The majority carriers produce the electric current in forward bias condition. So the electric current produced in forward bias condition is also known as majority current.

When the voltage is applied to the P-N junction diode in such a way that the positive terminal of the battery is connected to the n-type semiconductor and the negative terminal of the battery is connected to the p-type semiconductor, the diode is said to be reverse biased.

When this reverse bias voltage is applied to the P-N junction diode, a large number of free electrons (majority carriers) in the n-type semiconductor experience an attractive force from the positive terminal of the battery similarly a large number of holes (majority carriers) in the p-type semiconductor experience an attractive force from the negative terminal of the battery.

As a result, the free electrons (majority carriers) in the n-type semiconductor moves away from the P-N junction and attracted to the positive terminal of the battery similarly the holes (majority carriers) in the p-type semiconductor moves away from the P-N junction and attracted to the negative terminal of the battery.

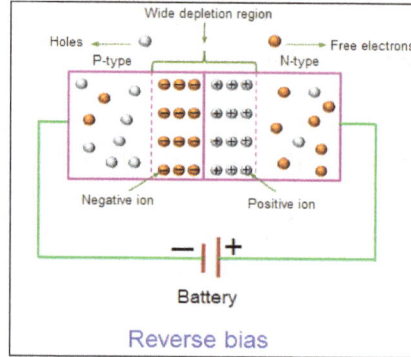

Reverse bias

Therefore, the electric current flow does not occur across the P-N junction. However, the minority carriers (free electrons) in the p-type semiconductor experience a repulsive force from the negative terminal of the battery similarly the minority carriers (holes) in the n-type semiconductor experience a repulsive force from the positive terminal of the battery.

As a result, the minority carriers free electrons in the p-type semiconductor and the minority carriers holes in the n-type semiconductor starts flowing across the junction. Thus, electric current is produced in the reverse bias diode due to the minority carriers. However, the electric current produced by the minority carriers is very small. So the minority carrier current in the reverse bias condition is neglected.

Thus, the P-N junction diode allows electric current in forward bias condition and blocks electric current in reverse bias condition. In simple words, a P-N junction diode allows electric current in only one direction. This unique property of the diode allows it to acts like a rectifier.

The forward bias and reverse bias voltage applied to the diode is nothing but a DC voltage. A DC voltage produces a current which always flows in one direction (either forward direction or backward direction).

But an AC voltage produces a current which always reverses its direction many times a second (forward to backward and backward to forward).

We have observed how a diode behaves when DC voltage (forward bias voltage and reverse bias voltage) is applied to it. Now let's take look at a P-N junction diode when AC voltage is applied to it.

The AC voltage or AC current is often represented by a sinusoidal waveform whereas the DC current is represented by a straight horizontal line.

In the sinusoidal waveform, the upper half cycle represents the positive half cycle and the lower half cycle represents the negative half cycle. The positive half cycle of the AC voltage is analogous to the forward bias DC voltage and the negative half cycle of the AC voltage is analogous to the reverse bias DC voltage.

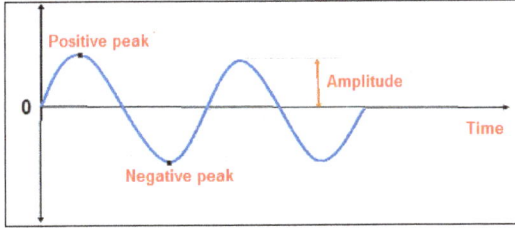

The alternating current starts from zero and grows to peak forward current or peak positive current. The positive peak of the sinusoidal waveform represents the maximum or peak forward current. After reaching the peak forward current, it starts decreasing and reaches to zero.

After a short period, the alternating current starts increasing in the reverse or negative direction and grows to peak reverse current or peak negative current. The negative peak of the sinusoidal waveform represents the maximum or peak reverse current. After reaching the peak reverse current, it starts decreasing and reaches to zero. Likewise, the alternating current continuously changes its direction in a short period.

When AC voltage or AC current is applied across the P-N junction diode, during the positive half cycle the diode is forward biased and allows electric current through it. However, when the AC current reverses its direction to negative half cycle, the diode is reverse biased and does not allow electric current through it. In simple words, during the positive half cycle, the diode allows current and during the negative half cycle, the diode blocks current. Thus, electric current flows through the diode only during the positive half cycle of the AC current.

This current which flows across the diode is nothing but a DC current. Thus, the P-N junction diode acts like a rectifier by converting the AC current into DC current.

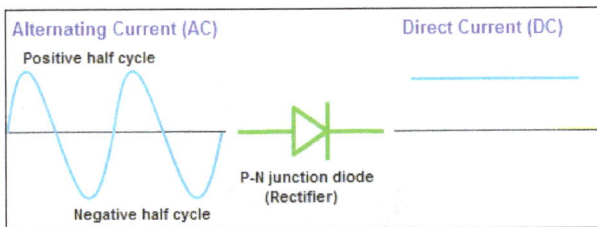

However, the DC current produced by a basic rectifier (half wave rectifier) is not a pure DC current. It is a pulsating DC current.

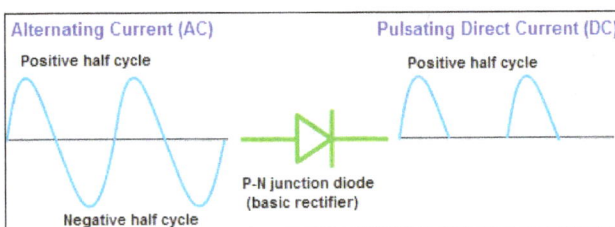

The pulsating direct current is a type of DC current whose value changes over a short period.

The pulsating DC current starts from zero and grows to the maximum forward current (peak level), and decreases to zero. However, the pulsating DC current does not change its direction periodically like AC current.

The pulsating DC current always flows in one direction like the pure DC current. However, the value of pulsating DC current or pulsating DC voltage slightly changes over a given period. The electric current produced by batteries, power supplies, and solar panels is a pure DC current.

By using the combination of components such as capacitors, inductors, and resistors in the circuit, we can achieve the smoothening of pulsating DC to pure DC.

Rectifier Practical Example

In our houses, almost all the electronic appliances use AC current. However, some electronic appliances such as laptops or notebook computers convert this AC current into DC current before they consume the power.

The AC adapter of the laptop connected to the AC source converts the high AC voltage or high AC current into low DC voltage or low DC current. This low DC current is supplied to the laptop battery and this is what we called laptop charging. However, the laptop will not turn on unless you manually turned it on by pressing the on button. When you press the laptop "power on" button, the laptop battery starts supplying DC current.

OUTSIDE : 4.8mm
INSIDE:1.7mm

POWER CORD
INCLUDED

We have forgotten an important step; how the AC adapters convert high AC voltage or high AC current into low DC voltage or low DC current. The AC adapters consist of all the essential components needed for AC to DC conversion. These components are a transformer, capacitor, and several diodes. Out of these components, the main key component is a diode which converts the alternating current into direct current.

- The transformer in the AC adapter reduces the high AC voltage to a low AC voltage.

- The rectifier (made up of diodes) converts this low AC voltage or AC current into low DC voltage or DC current. However, the converted current is not pure DC current. It is a pulsating DC current.

- The capacitor filters this pulsating DC current to pure DC current.

SINGLE PHASE UNCONTROLLED RECTIFIERS

Half Wave Rectifier

A half wave rectifier is defined as a type of rectifier that only allows one half-cycle of an AC voltage waveform to pass, blocking the other half-cycle. Half-wave rectifiers are used to convert AC voltage to DC voltage, and only require a single diode to construct.

A rectifier is a device that converts alternating current (AC) to direct current (DC). It is done by using a diode or a group of diodes. Half wave rectifiers use one diode, while a full wave rectifier uses multiple diodes.

The working of a half wave rectifier takes advantage of the fact that diodes only allow current to flow in one direction.

Half Wave Rectifier Theory

A half wave rectifier is the simplest form of rectifier available.

The diagram below illustrates the basic principle of a half-wave rectifier. When a standard AC waveform is passed through a half-wave rectifier, only half of the AC waveform remains. Half-wave rectifiers only allow one half-cycle (positive or negative half-cycle) of the AC voltage through and will block the other half-cycle on the DC side, as seen below.

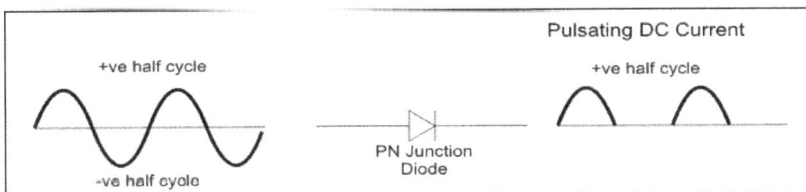

Only one diode is required to construct a half-wave rectifier. In essence, this is all that the half-wave rectifier is doing.

Since DC systems are designed to have current flowing in a single direction (and constant voltage), putting an AC waveform with positive and negative cycles through a DC device can have destructive (and dangerous) consequences. So we use half-wave rectifiers to convert the AC input power into DC output power.

But the diode is only part of it – a complete half-wave rectifier circuit consists of 3 main parts:

1. A transformer,

2. A resistive load,

3. A diode.

A half wave rectifier circuit diagram looks like this:

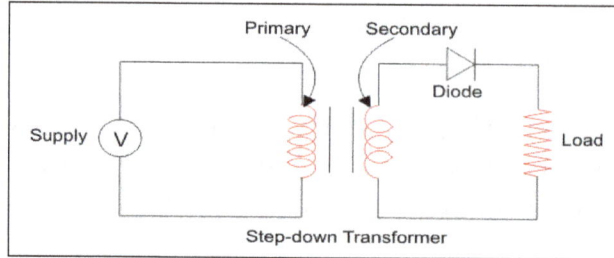

We'll now go through the process of how a half-wave rectifier converts an AC voltage to a DC output.

First, a high AC voltage is applied to the to the primary side of the step-down transformer and we will get a low voltage at the secondary winding which will be applied to the diode.

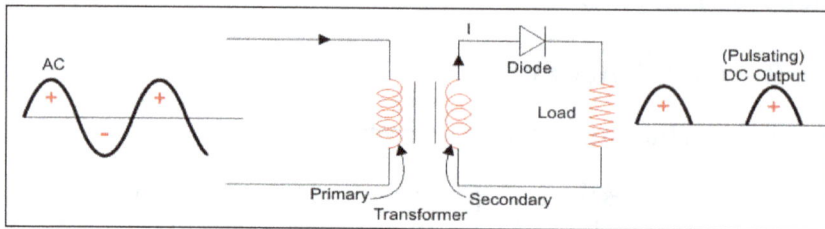

During the positive half cycle of the AC voltage, the diode will be forward biased and the current flows through the diode. During the negative half cycle of the AC voltage, the diode will be reverse biased and the flow of current will be blocked. The final output voltage waveform on the secondary side (DC) is shown in figure above.

This can be confusing on first glance – so let's dig into the theory of this a bit more.

We'll focus on the secondary side of the circuit. If we replace the secondary transformer coils with a source voltage, we can simplify the circuit diagram of the half-wave rectifier as:

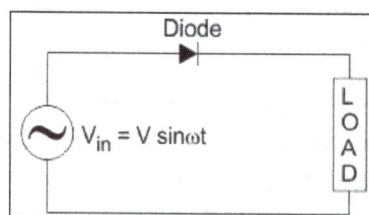

Now we don't have the transformer part of the circuit distracting us.

For the positive half cycle of the AC source voltage, the equivalent circuit effectively becomes:

This is because the diode is forward biased, and is hence allowing current to pass through. So we have a closed circuit.

But for the negative half cycle of the AC source voltage, the equivalent circuit becomes:

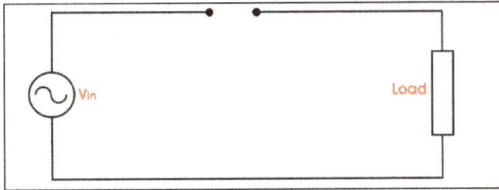

Because the diode is now in reverse bias mode, no current is able to pass through it. As such, we now have an open circuit. Since current can not flow through to the load during this time, the output voltage is equal to zero.

This all happens very quickly – since an AC waveform will oscillate between positive and negative many times each second (depending on the frequency).

Here's what the half wave rectifier waveform looks like on the input side (V_{in}), and what it looks like on the output side (V_{out}) after rectification (i.e. conversion from AC to DC):

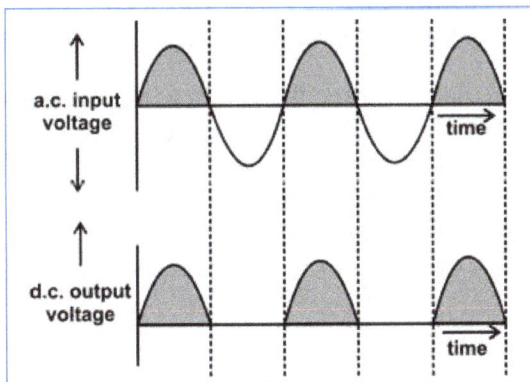

The graph above actually shows a positive half wave rectifier. This is a half-wave rectifier which only allows the positive half-cycles through the diode, and blocks the negative half-cycle.

The voltage waveform before and after a positive half wave rectifier is shown in figure below:

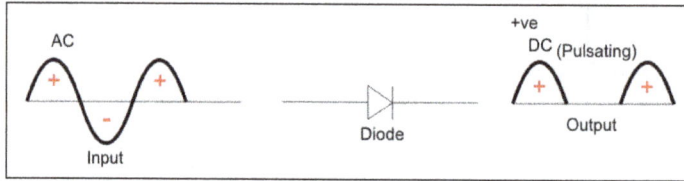

Conversely, a negative half-wave rectifier will only allow negative half-cycles through the diode and will block the positive half-cycle. The only difference between a posive- and negative half wave rectifier is the direction of the diode.

As you can see in figure below, the diode is now in the opposite direction. Hence the diode will now be forward biased only when the AC waveform is in its negative half cycle.

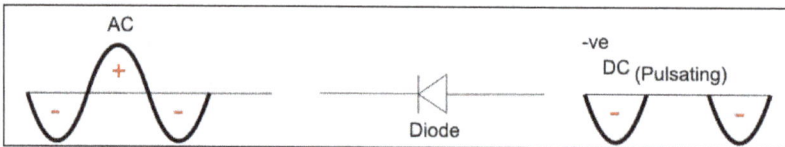

Half Wave Rectifier Capacitor Filter

The output waveform we have obtained from the theory above is a pulsating DC waveform. This is what is obtained when using a half wave rectifier without a filter.

Filters are components used to convert (smoothen) pulsating DC waveforms into constant DC waveforms. They achieve this by suppressing the DC ripples in the waveform.

Although half-wave rectifiers without filters are theoretically possible, they can't be used for any practical applications. As DC equipment requires a constant waveform, we need to 'smooth out' this pulsating waveform for it to be any use in the real world.

This is why in reality we use half wave rectifiers with a filter. A capacitor or an inductorcan be used as a filter – but half wave rectifier with capacitor filter is most commonly used.

The circuit diagram below shows how a capacitive filter is can be used to smoothen out a pulsating DC waveform into a constant DC waveform.

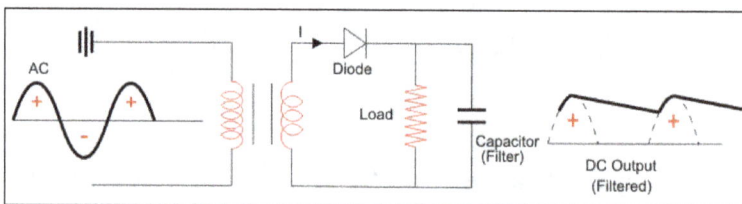

Ripple Factor of Half Wave Rectifier

'Ripple' is the unwanted AC component remaining when converting the AC voltage waveform into a DC waveform. Even though we try out best to remove all AC components, there is still some small amount left on the output side which pulsates the DC waveform. This undesirable AC component is called 'ripple'.

To quantify how well the half-wave rectifier can convert the AC voltage into DC voltage, we use what is known as the ripple factor (represented by γ or r). The ripple factor is the ratio between the RMS value of the AC voltage (on the input side) and the DC voltage (on the output side) of the rectifier.

The formula for ripple factor is:

$$\gamma = \sqrt{\left(\frac{V_{rms}}{V_{DC}}\right)^2 - 1}$$

Which can also be rearranged to equal:

$$Ripple\ facto(r) = \frac{\left(I_{rms}^2 - I_{dc}^2\right)}{I_{dc}} = 1.21$$

The ripple factor of half wave rectifier is equal to 1.21 (i.e. γ = 1.21).

Note that for us to construct a good rectifier, we want to keep the ripple factor as low as possible. This is why we use capacitors and inductors as filters to reduce the ripples in the circuit.

Efficiency of Half Wave Rectifier

Rectifier efficiency (η) is the ratio between the output DC power and the input AC power. The formula for the efficieny is equal to:

$$\eta = \frac{P_{dc}}{P_{ac}}$$

The efficiency of a half wave rectifier is equal to 40.6% (i.e. η_{max} = 40.6%).

RMS Value of Half Wave Rectifier

To derive the RMS value of half wave rectifier, we need to calculate the current across the load. If the instantaneous load current is equal to $i_L = I_m \sin \omega t$, then the average of load current (I_{DC}) is equal to:

$$I_{dc} = \frac{1}{2\pi} \int_0^\pi I_m \sin \omega t = \frac{I_m}{\pi}$$

Where I_m is equal to the peak instantaneous current across the load (I_{max}). Hence the output DC current (I_{DC}) obtained across the load is:

$$I_{dc} = \frac{I_{max}}{\pi}, \; where \; I_{max} = maximum \; amplitude \; of \; dc \; current$$

For a half-wave rectifier, the RMS load current (I_{rms}) is equal to the average current (I_{DC}) multiple by $\pi/2$. Hence the RMS value of the load current (I_{rms}) for a half wave rectifier is:

$$I_{rms} = \frac{I_m}{4}$$

Where, $I_m = I_{max}$ which is equal to the peak instantaneous current across the load.

Peak Inverse Voltage of Half Wave Rectifier

Peak Inverse Voltage (PIV) is the maximum voltage that the diode can withstand during reverse bias condition. If a voltage is applied more than the PIV, the diode will be destroyed.

Form Factor of Half Wave Rectifier

Form factor (F.F) is the ratio between RMS value and average value, as shown in the formula below:

$$F.F = \frac{RMS \, value}{Avarage \; value}$$

The form factor of a half wave rectifier is equal to 1.57 (i.e. F.F= 1.57).

Output DC Voltage

The output voltage (V_{DC}) across the load resistor is denoted by:

$$V_{DC} = \frac{V_{S\,max}}{\pi}, \; where \; V_{S\,max} = maximum \, amplitude \; of \; secondary \; voltage.$$

Applications of Half Wave Rectifier

Half wave rectifiers are not as commonly used as full-wave rectifiers. Despite this, they still have some uses:

- For rectification applications.

- For signal demodulation applications.

- For signal peak applications.

Advantages of Half Wave Rectifier

The main advantage of half-wave rectifiers is in their simplicity. As they don't require as many components, they are simpler and cheaper to setup and construct.

As such, the main advantages of half-wave rectifiers are:

- Simple (lower number of components).

- Cheaper up front cost (as their is less equipment. Although there is a higher cost over time due to increased power losses).

Disadvantages of Half Wave Rectifier

The disadvantages of half-wave rectifiers are:

- They only allow a half-cycle through per sinewave, and the other half-cycle is wasted. This leads to power loss.

- They produces a low output voltage.

- The output current we obtain is not purely DC, and it still contains a lot of ripple (i.e. it has a high ripple factor).

Three Phase Half Wave Rectifier

All of the theory above has dealt with a single phase half wave rectifier. Although the principle of a 3 phase half wave rectifier is the same, the characteristics are different. The waveform, ripple factor, efficiency, and RMS output values are not the same.

The three phase half wave rectifier is used for conversion of three-phase AC power to DC power. Here the switches are diodes, and hence they are uncontrolled switches. That is to say, there is no way of controlling the on and off times of these switches.

The 3 phase half wave diode rectifier is generally constructed with a three-phase supply connected to a three-phase transformer where the secondary winding of the transformer is always connected via star connection. This is because the neutral point is required to connect the load back to the transformer secondary windings, providing a return path for the flow of power.

A typical configuration of a three-phase half wave rectifier supplying to a purely resistive load is shown below. Here, each phase of the transformer is considered as an individual alternating source. The simulation and measurement of voltages are as shown in the circuit below. Here we have connected an individual voltmeter across each source as well as across the load.

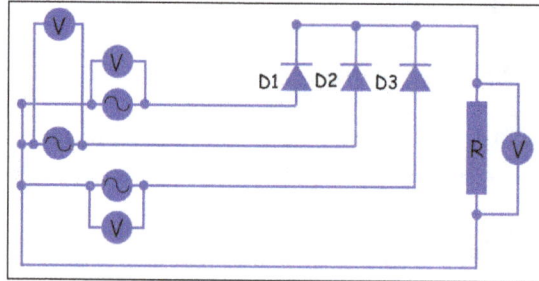

The three-phase voltages are shown below:

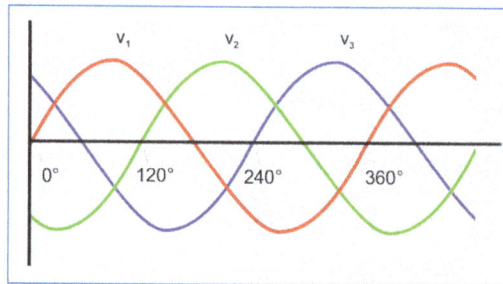

The voltage across the resistive load is shown below. The voltage is shown in black.

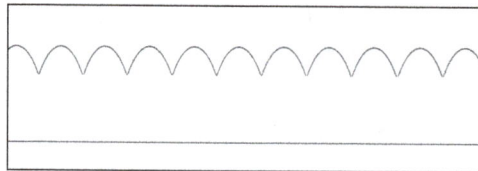

So we can see from the above figure that the diode D1 conducts when the R phase has a value of thevoltage that is higher than the value of the voltage of the other two phases,and this condition begins when the R phase is at a 30° and repeats after every complete cycle. That is to say, the next time diode DI begins to conduct is at 390°. Diode D2 takes over conduction from D1 which stops conducting at angle 150° because at this instant the value of voltage in B phase becomes higher than the voltages in the other two phases. So each diode conducts for an angle of 150° − 30° = 120°.

Here, the waveform of the resulting DC voltage signal is not purely DC as it is not flat,but rather it contains a ripple. And the frequency of the ripple is 3 × 50 = 150 Hz.

The average of the output voltage across the resistive load is given by,

$$V_o = \frac{1}{\frac{2\pi}{3}} \int_{\frac{\pi}{6}}^{\frac{5\pi}{6}} V_{mphase} \sin(\omega t)\, d(\omega t)$$

$$= \frac{3\sqrt{6}}{2\pi} V_{phase} = 1.17\, V_{phase} = 0.827\, V_{mphase} = \frac{3}{2\pi} V_{m\,line}$$

Where,

$$V_{mphase} = \sqrt{2}\, V_{phase}$$

$$V_{m\,line} = \sqrt{3} V_{mphase} = \sqrt{6} V_{phase}$$

The RMS value of the output voltage is given by,

$$V_{o_{rms}} = \left[\frac{1}{\frac{2\pi}{3}} \int_{\frac{\pi}{6}}^{\frac{5\pi}{6}} \left(V_{mphase} \sin(\omega t) \right)^2 d(\omega t) \right]^{\frac{1}{2}}$$

$$= 0.84068 V_{mphase}$$

The ripple voltage is equal to,

$$V_r = \sqrt{V_{o_{rms}}^2 - V_0^2}$$

$$= \sqrt{0.84068^2 - 0.827^2}\, V_{mphase}$$

$$= 0.151\, V_{mphase}$$

And the voltage ripple factor is equal to,

$$\frac{V_r}{V_o} = \frac{0.151}{0.827} = 0.1826 = 18.26\%$$

The equation above shows that the voltage ripple is significant. This is undesirable as this leads to unnecessary power loss.

DC output power,

$$P_o = V_o I_o = \frac{3\sqrt{3}}{2\pi} V_{mphase} \frac{3\sqrt{3}}{2\pi} I_{mphase}$$

$$= 0.684\, V_{mphase}\, I_{mphase}$$

AC input power,

$$P_i = V_{or} \, I_{or} = (0.84068)^2 \, V_{mphase} \, I_{mphase}$$
$$= 0.7067 \, V_{mphase} \, I_{mphase}$$

Efficiency,

$$\eta = \frac{P_o}{P_i} = \frac{0.684 \, V_{mphase} \, I_{mphase}}{0.7067 \, V_{mphase} \, I_{mphase}} = 0.968 = 96.8\%$$

Even though the efficiency of the 3 phase half-wave rectifier is seemingly high, it is still less than the efficiency provided by a 3 phase full wave diode rectifier. Although three phase half wave rectifiers are cheaper, this cost saving is insignificant compared to the money lost in their higher power losses. As such, three-phase half-wave rectifiers are not commonly used in industry.

Full Wave Rectifier

A full wave rectifier converts both halves of each cycle of an alternating wave (AC signal) into pulsating DC signal.

We can further classify full wave rectifiers into:

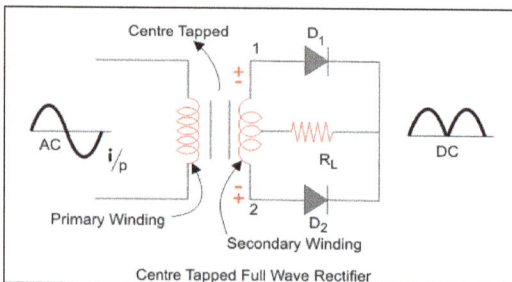

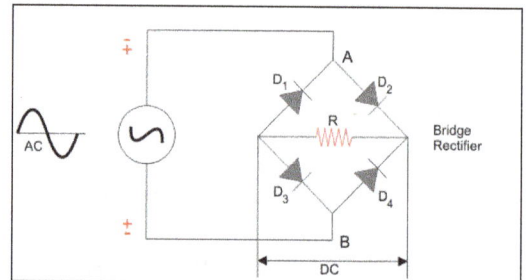

- Centre-tapped Full Wave Rectifier,

- Full Wave Bridge Rectifier.

Centre-tapped Full Wave Rectifier

Construction of Centre-tapped Full Wave Rectifier

A centre-tapped full wave rectifier system consists of:

- Centre-tapped Transformer,

- Two Diodes,

- Resistive Load.

Centre-tapped Transformer: It is a normal transformer with one slight modification. It has an addition wire connected to the exact centre of the secondary winding. This type of construction divides the AC voltage into two equal and opposite voltages namely +Ve voltage (V_a) and -Ve voltage (V_b). The total output voltage is:

$$V = V_a + V_b$$

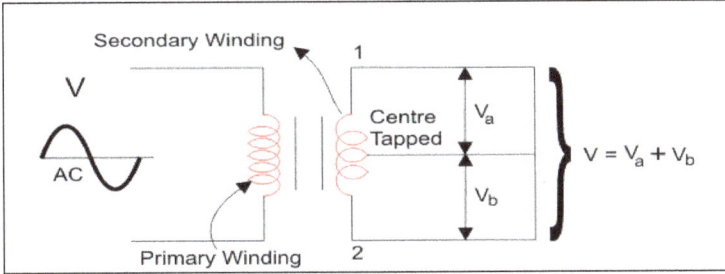

The circuit diagram is as follows:

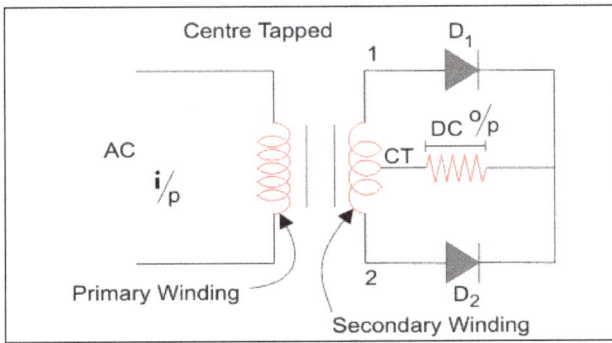

Working of Centre-tapped Full Wave Rectifier

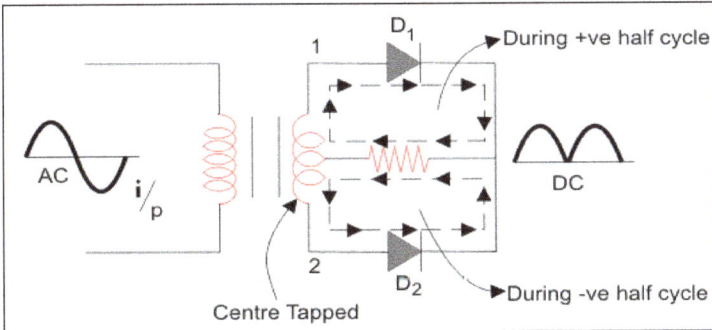

We apply an AC voltage to the input transformer. During the positive half-cycle of the AC voltage, terminal 1 will be positive, centre-tap will be at zero potential and terminal 2 will be negative potential. This will lead to forward bias in diode D_1 and cause current to flow through it. During this time, diode D_2 is in reverse bias and will block current through it.

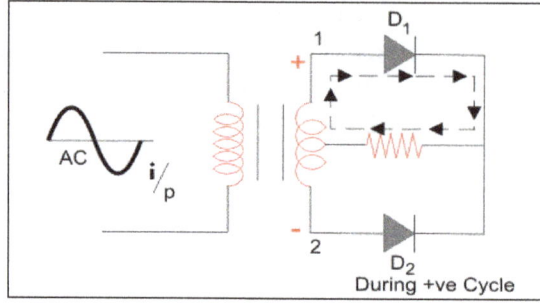

During +ve Cycle

During the negative half-cycle of the input AC voltage, terminal 2 will become positive with relative to terminal 2 and centre-tap. This will lead to forward bias in diode D_2 and cause current to flow through it. During this time, diode D_1 is in reverse bias and will block current through it.

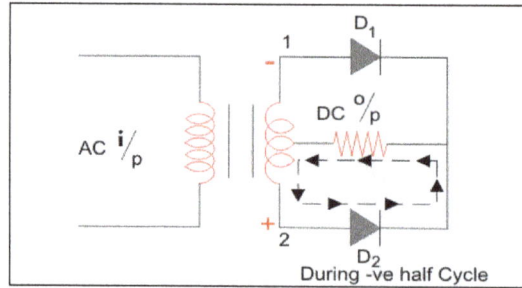

During -ve half Cycle

During the positive cycle, diode D_1 conducts and during negative cycle diode D_2 conducts and during positive cycle. As a result, both half-cycles are allowed to pass through. The average output DC voltage here is almost twice of the DC output voltage of a half-wave rectifier.

Output Waveforms

Filter Circuit

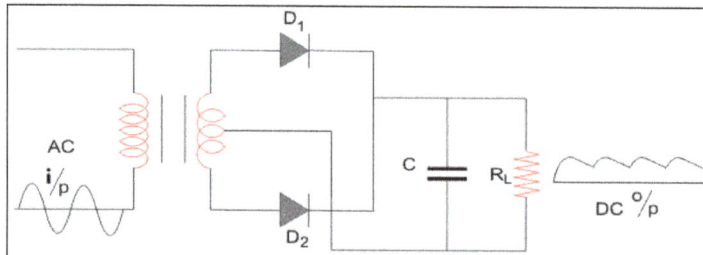

We get a pulsating DC voltage with a lot of ripples as the output of the centre-tapped full wave rectifier. We cannot use this pulsating for practical applications. So, to convert the pulsating DC voltage to pure DC voltage, we use a filter circuit as shown above. Here we place a capacitor across the load. The working of the capacitive filter circuit is to short the ripples and block the DC component so that it flows through another path and is available across the load. During the positive half-wave, the diode D_1 starts conducting. The capacitor is uncharged, and when we apply an input AC voltage which happens to be more than the capacitor voltage, it charges the capacitor immediately to the maximum value of the input voltage. At this point, the supply voltage is equal to capacitor voltage.

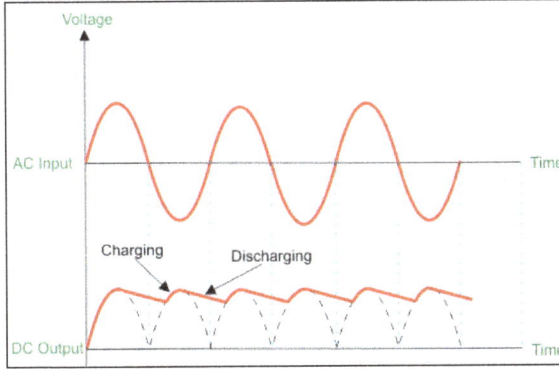

When the applied AC voltage starts decreasing and less than the capacitor, the capacitor starts discharging slowly but this is slower when compared to the charging of the capacitorand it does not get enough time to discharge entirely and the charging starts again. So around half of the charge present in the capacitor gets discharged. During the negative cycle, the diode D_2 starts conducting, and the above process happens again. This will cause the current to flow through the same direction across the load.

Full Wave Bridge Rectifier

Construction of Full Wave Bridge Rectifier

A full wave bridge rectifier is a type of rectifier which will use four diodes or more than that in a bridge formation. A full wave bridge rectifier system consists of:

1. Four Diodes,

2. Resistive Load.

We use the diodes namely A, B, C and D which form a bridge circuit. The circuit diagram is as follows:

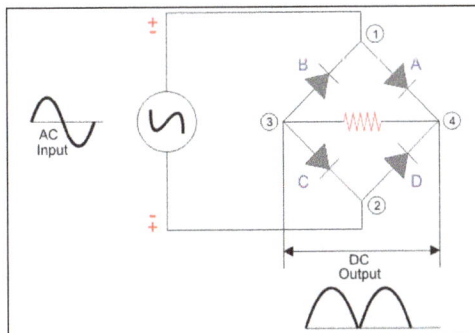

Principle of Full Wave Bridge Rectifier

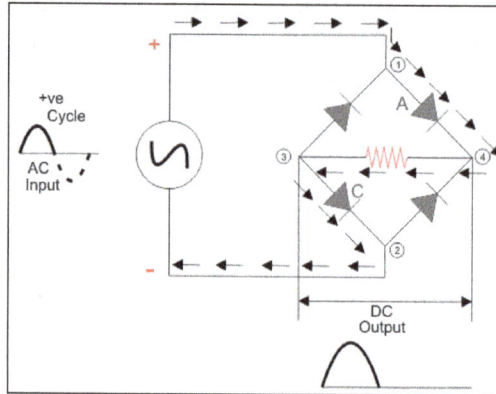

We apply an AC across the bridge. During the positive half-cycle, the terminal 1 becomes positive, and terminal 2 becomes negative. This will cause the diodes A and C to become forward-biased, and the current will flow through it. Meanwhile diodes B and D will become reverse-biased and block current through them. The current will flow from 1 to 4 to 3 to 2.

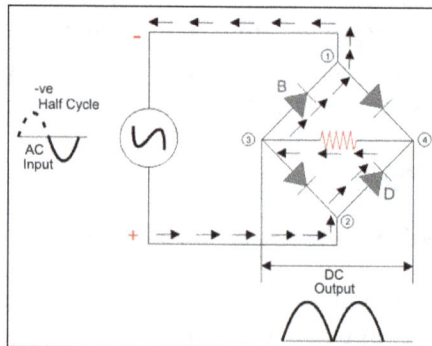

During the negative half-cycle, the terminal 1 will become negative, and terminal 2 will become positive. This will cause the diodes B and D to become forward-biased and will allow current through them. At the same time, diodes A and C will be reverse-biased and will block the current through them. The current will flow from 2 to 4 to 3 to 1.

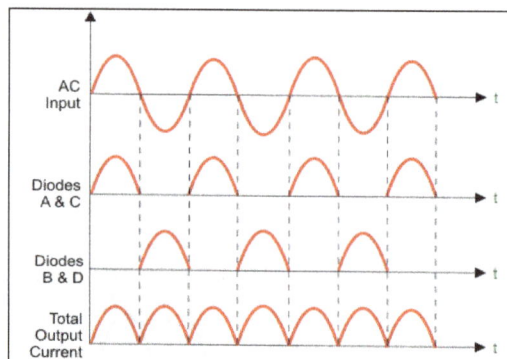

Filter Circuit

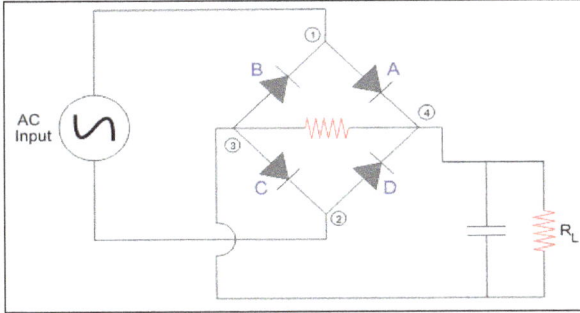

We get a pulsating DC voltage with a lot of ripples as the output of the full wave bridge rectifier. We can not use this voltage for practical applications. So, to convert the pulsating DC voltage to pure DC voltage, we use a filter circuit as shown above. Here we place a capacitor across the load. The working of the capacitive filter circuit is to short the ripples and block the DC component so that it flows through another path and that is through the load. During the half-wave, the diodes A and C conduct. It charges the capacitor immediately to the maximum value of the input voltage. When the rectified pulsating voltage starts decreasing and less than the capacitor voltage, the capacitor starts discharging and supplies current to the load. This discharging is slower when compared to the charging of the capacitor, and it does not get enough time to discharge entirely and the charging starts again in next pulse of the rectified voltage waveform. So around half of the charge present in the capacitor gets discharged. During the negative cycle, the diodes B and D start conducting, and the above process happens again. This causes, the current continues to flow through the same direction across the load.

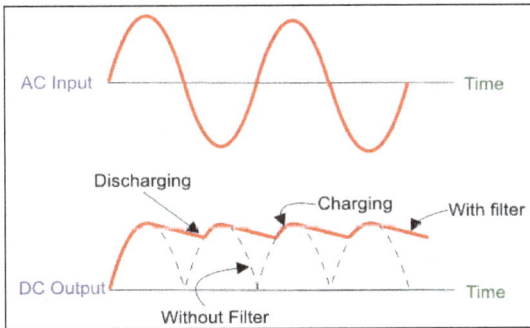

Characteristics of a Full-wave Rectifier

Ripple Factor (γ)

$$\gamma = \sqrt{\left(\frac{V_{rms}}{V_{dc}}\right)^2 - 1}$$

The output we will get from the rectifier will consist of both AC and DC components. The

AC components are undesirable to us and will cause pulsations in the output. This unwanted AC components are called Ripple. The expression ripple factor is given above where V_{rms} is the RMS value of the AC component and V_{dc} is the DC component in the rectifier.

For centre-tapped full-wave rectifier, we obtain $\gamma = 0.48$.

For us to construct a good rectifier, we need to keep the ripple factor as minimum as possible. We can use capacitors or inductors to reduce the ripples in the circuit. Rectifier Efficiency (η).

Rectifier efficiency is the ratio between the output DC power and the input AC power:

$$\eta = \frac{P_{dc}}{P_{ac}}$$

For centre-tapped full-wave rectifier, $\eta_{max} = 81.2\%$.

Form Factor (F.F): The form factor is the ratio between RMS value and average value.

$$FF = \frac{RMS\,value}{Average\,value}$$

For centre-tapped full wave rectifier, FF = 1.11.

Advantages of Full Wave Rectifiers

- Full wave rectifiers have higher rectifying efficiency than half-wave rectifiers. This means that they convert AC to DC more efficiently.

- They have low power loss because no voltage signal is wasted in the rectification process.

- The output voltage of centre-tapped full wave rectifier has lower ripples than a halfwave rectifiers.

Disadvantages of Full Wave Rectifiers

- The centre-tapped rectifier is more expensive than half-wave rectifier and tends to occupy a lot of space.

SINGLE PHASE CONTROLLED RECTIFIERS

Single Phase Half Controlled Rectifiers

Single Phase Half Wave Controlled Rectifier, is a rectifier circuit which converts AC input into DC output only for positive half cycle of the AC input supply. The word "controlled" means that, we can change the starting point of load current by controlling the firing angle of SCR. These words might seem a lot technical. But firing of SCR simply means, the SCR turn ON at certain point of time when it is forward biased.

A Single Phase Half Wave Controlled Rectifier circuit consists of SCR / thyristor, an AC voltage source and load. The load may be purely resistive, Inductive or a combination of resistance and inductance. For simplicity, we will consider a resistive load. A simple circuit diagram of Single Phase Half Wave Controlled Rectifier is shown in figure below:

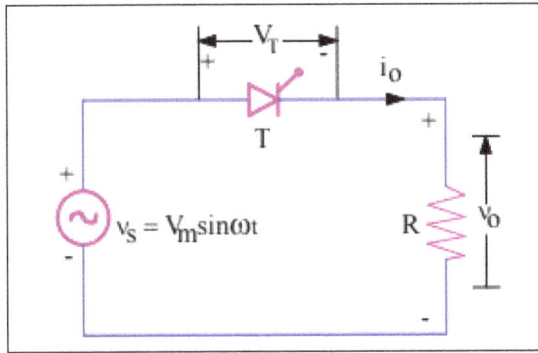

v_o = Load output voltage,

i_o = Load current,

V_T = Voltage across the thyristor T.

Following points must be kept in mind while discussing controlled rectifier:

- The necessary condition for turn ON of SCR is that, it should be forward biased and gate signal must be applied. In other words, an SCR will only get turned ON when it is forward biased and fired or gated.

- SCR will only turn off when current through it reaches below holding current and reverse voltage is applied for a time period more than the SCR turn off time.

Well, let us go ahead with the above points in mind. Let us assume that thyristor T is fired at a firing angle of α. This means when wt = α, gate signal will be applied and SCR will start conducting.

Thyristor T is forward biased for the positive half cycle of supply voltage. The load output voltage is zero till SCR is fired. Once SCR is fired at an angle of α, SCR starts conducting. But as soon as the supply voltage becomes zero at $\omega t = \pi$, the load current will become zero and after $\omega t = \pi$, SCR is reversed biased. Thus thyristor T will turn off at $\omega t = \pi$ and will remain in OFF condition till it is fired again at $\omega t = (2\pi + \alpha)$.

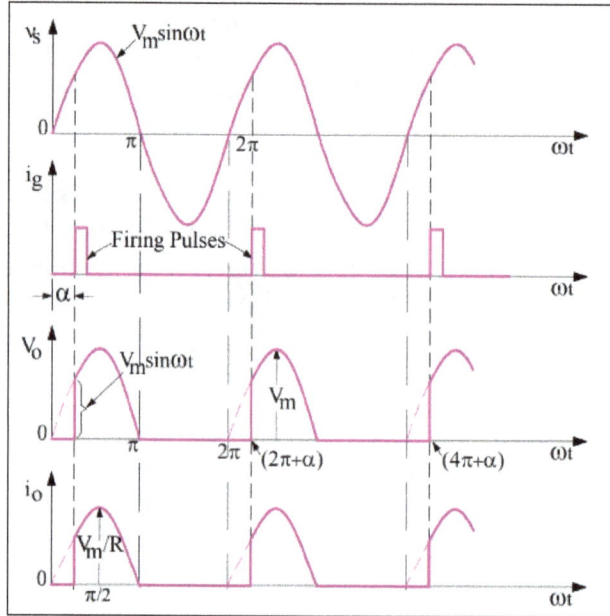

Therefore, the load output voltage and current for one complete cycle of input supply voltage may be written as:

$v_0 = V_m \sin \omega t$ for $\alpha \le \omega t \le \pi$

$i_0 = V_m \sin \omega t / R$ for for $\alpha \le \omega t \le \pi$

Calculation of Average Load Output Voltage

As we know that, average value of any function f(x) cab be calculated using the formula:

$$\text{Average Value} = (1/T)\int_0^T f(x)\,dx$$

Let us now calculate the average value of output voltage for Single Phase Half Wave Controlled Rectifier.

Average Value of Load output voltage:

$$= (1/2\pi)\int_0^{2\pi} V_m \sin \omega t \, d(\omega t)$$

$$= (1/2\pi)\int_0^{\alpha} V_m \sin \omega t \, d(\omega t) + \int_{\alpha}^{\pi} V_m \sin \omega t \, d(\omega t) + \int_{\pi}^{2\pi} V_m \sin \omega t \, d(\omega t)$$

Since the value of load output voltage is zero from $0 \le \omega t \le \alpha$ and $\pi < \omega t < 2\pi$, therfore

$$= (1/2\pi)\int_0^{\pi} Vm \sin \omega t \, d(\omega t)$$

$$= (Vm/2\pi) \int_{\alpha}^{\pi} Sin\ \omega t\ d(\omega t)$$

$$= \left(\frac{Vm}{2\pi}\right)[1+Cos\alpha]$$

For single phase half wave controlled rectifier:

Avarage value of Load output Voltage $= \left(\frac{Vm}{2\pi}\right)[1+Cos\ \alpha]$

From the expression of average output voltage, it can be seen that, by changing firing angle α, we can change the average output voltage. The average output voltage is maximum when firing angle is zero and it is minimum when firing angle α = π. This is the reason, it is called phase controlled rectifier.

Average load current for Single Phase Half Wave Controlled Rectifier can easily be calculated by dividing the average load output voltage by load resistance R.

Let us now calculate the root mean square (rms) value of load voltage.

$$RMS\ Value = \sqrt{(1/T)\int_{0}^{T}\left[f(x)\right]^2 dx}$$

RMS Value of Load output Voltage,

$$= \sqrt{(1/2\pi)\int_{0}^{2\pi}\left[Vm\sin\omega t\right]^2 d(\omega t)}$$

$$= \sqrt{(Vm/4\pi)\int_{0}^{2\pi}\left[Vm\sin\omega t\right]^2 d(\omega t)}$$

$$= \sqrt{(Vm/4\pi)\int_{0}^{2\pi}\left[1-Cos2\ \omega t\right]d(\omega t)}$$

Since the value of load output voltage is zero from $0 \le \omega t \le \alpha$ and $\pi < \omega t < 2\pi$, therfore

$$= \sqrt{(Vm/4\pi)\int_{0}^{\pi}\left[1-Cos2\ \omega t\right]d(\omega t)}$$

$$= \left(\frac{Vm}{2\sqrt{\pi}}\right)\sqrt{(\pi-\alpha)-(1/2)Sin\ 2\alpha}$$

RMS Value of Load Output Voltage,

$$= \left(\frac{Vm}{2\sqrt{\pi}}\right)\sqrt{(\pi-\alpha)-(1/2)Sin\ 2\alpha}$$

RMS value of load current can be calculated by dividing the rms load voltage by resistance R. This means,

RMS Load Current I_{orms} = RMS Load Voltage/R.

Input volt ampere can be calculated as:

Input Volt Ampere = RMS Supply Voltage x RMS Load Current

$$= V_s \times I_{orms}.$$

Single Phase Fully Controlled Rectifier

Figure below shows the Single phase Full Wave Controlled Rectifiers with R load.

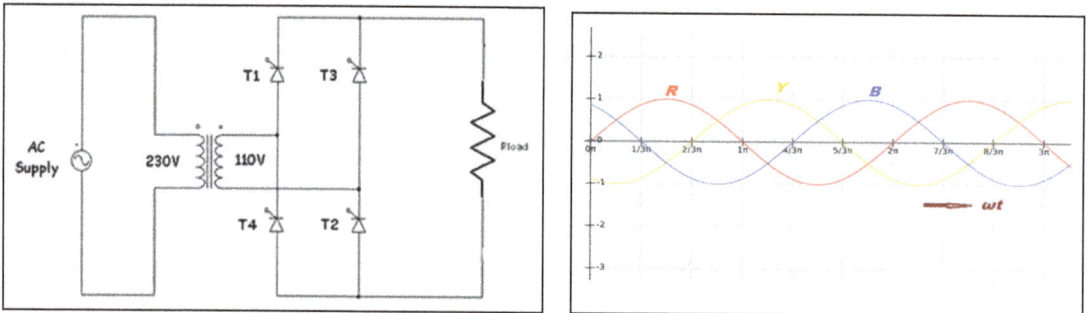

- The single phase fully controlled rectifier allows conversion of single phase AC into DC. Normally this is used in various applications such as battery charging, speed control of DC motors and front end of UPS (Uninterruptible Power Supply) and SMPS (Switched Mode Power Supply).

- All four devices used are thyristors. The turn-on instants of these devices are dependent on the firing signals that are given. Turn-off happens when the current through the device reaches zero and it is reverse biased at least for duration equal to the turn-off time of the device specified in the data sheet.

- In positive half cycle thyristors T1 & T2 are fired at an angle α.

- When T1 & T2 conducts.

  Vo=Vs.

  Io=Is=Vo/R=Vs/R.

- In negative half cycle of input voltage, SCR's T3 & T4 are triggered at an angle of (π+α).

- Here output current & supply current are in opposite direction,

  ∴ is = - io.

  T3 & T4 becomes off at 2π.

Single Phase Full Wave Controlled Rectifier with 'RL' load

Figure below shows Single phase Full Wave Controlled Rectifiers with RL load.

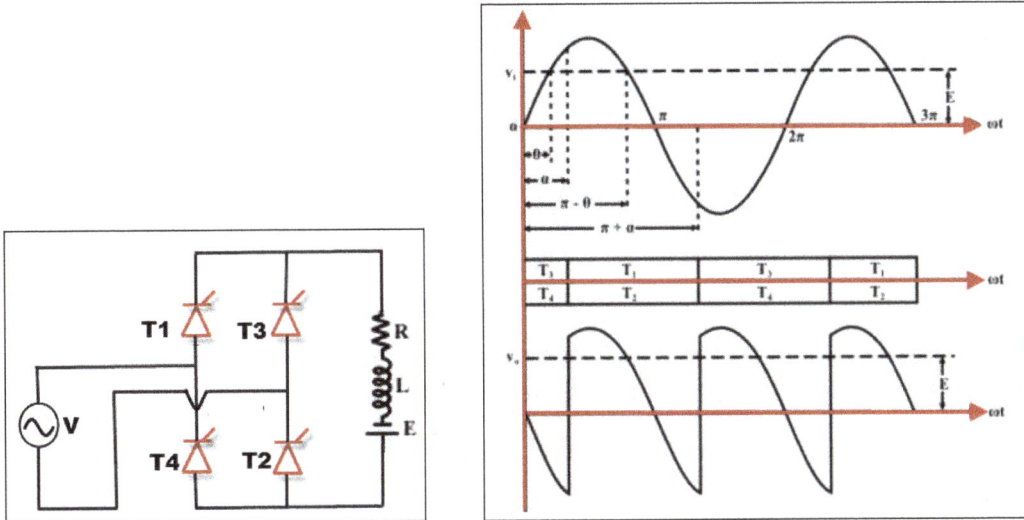

Operation of this Mode can be Divided between Four Modes

Mode 1 (α to π):

- In positive half cycle of applied ac signal, SCR's T1 & T2 are forward bias & can be turned on at an angle α.

- Load voltage is equal to positive instantaneous ac supply voltage. The load current is positive, ripple free, constant and equal to Io.

- Due to positive polarity of load voltage & load current, load inductance will store energy.

Mode 2 (π to $\pi + \alpha$):

- At wt=π, input supply is equal to zero & after π it becomes negative. But inductance opposes any change through it.

- In order to maintain a constant load current & also in same direction. A self inducedemf appears across 'L' as shown.

- Due to this induced voltage, SCR's T1 & T2 are forward bais in spite the negative supply voltage.

- The load voltage is negative & equal to instantaneous ac supply voltage whereas load current is positive.

- Thus, load acts as source & stored energy in inductance is returned back to the ac supply.

Mode 3 (π + α to 2π):

- At wt=π+α SCR's T3 & T4 are turned on & T1, T2 are reversed bias.
- Thus , process of conduction is transferred from T1,T2 to T3,T4.
- Load voltage again becomes positive & energy is stored in inductor.
- T3, T4 conduct in negative half cycle from (π+α) to 2π.
- With positive load voltage & load current energy gets stored.

Mode 4 (2π to 2π+α):

- At wt=2π, input voltage passes through zero.
- Inductive load will try to oppose any change in current if in order to maintain load current constant & in the same direction.
- Induced emf is positive & maintains conducting SCR's T3 & T4 with reverse polarity also.
- Thus VL is negative & equal to instantaneous ac supply voltage. Whereas load current continues to be positive.
- Thus load acts as source & stored energy in inductance is returned back to ac supply.
- At wt=α or 2π+α, T3 & T4 are commutated and T1,T2 are turned on.

THREE PHASE RECTIFIERS

A 3 Phase rectifier is a device which rectifies the input AC voltage with the use of 3 phase transformer and 3 diodes which are connected to each of the three phases of transformer secondary winding.

Significance of 3 Phase Rectifier

A single phase rectifier also rectifies i.e. converts AC supply to DC supply but uses only single phase of transformer secondary coil for the conversion. And the diodes are connected to the secondary winding of single phase transformer.

The drawback of this arrangement is high ripple factor. In case of half wave rectifier the ripple factor is 1.21 and in case of full wave rectifier the ripple factor is 0.482. In both the cases the value of ripple factor cannot be neglected. While in case of half wave rectifier the value is quite large but in full wave rectifier too the value of rectifier is significantly large.

Thus, in such types of arrangement we need smoothing circuit in order to remove these ripples. These ripples are the AC components in the DC voltage. This is called pulsating DC voltage. If this pulsating DC voltage is used in several applications it lead to poor performance of the device. Thus, the Smoothing circuit is used, filter works as a smoothing circuit for rectifier system.

But after this smoothing process the rectifier voltage falls to zero at some point. Therefore, if in place of single phase transformer we use three phase transformer the ripple factor can be reduced up to a large extent. One of the significant advantage of three phase transformer is that the rectified voltage do not falls to zero even when no smoothing arrangement is used.

Half-wave Three-phase Rectification

The anode of each diode is connected to one phase of the voltage supply with the cathodes of all three diodes connected together to the same positive point, effectively creating a diode-"OR" type arrangement. This common point becomes the positive (+) terminal for the load while the negative (-) terminal of the load is connected to the neutral (N) of the supply.

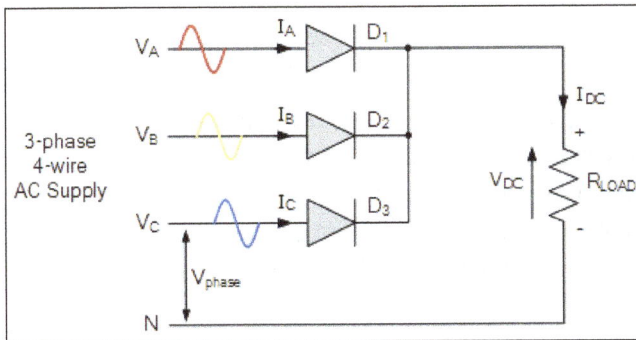

Assuming a phase rotation of Red-Yellow-Blue ($V_A – V_B – V_C$) and the red phase (V_A) starts at 0°. The first diode to conduct will be diode 1 (D^1) as it will have a more positive voltage at its anode than diodes D_2 or D_3. Thus diode D_1 conducts for the positive half-cycle of V_A while D_2 and D_3 are in their reverse-biased state. The neutral wire provides a return path for the load current back to the supply.

120 electrical degrees later, diode 2 (D_2) starts to conduct for the positive half-cycle of V_B (yellow phase). Now its anode becomes more positive than diodes D_1 and D_3 which are both "OFF" because they are reversed-biased. Similarly, 120° later V_C (blue phase) starts to increase turning "ON" diode 3 (D_3) as its anode becomes more positive, thus turning "OFF" diodes D_1 and D_2.

Then we can see that for three-phase rectification, whichever diode has a more positive voltage at its anode compared to the other two diodes it will automatically start to conduct, thereby giving a conduction pattern of: $D_1 D_2 D_3$ as shown.

Half-wave Three-phase Rectifier Conduction Waveform

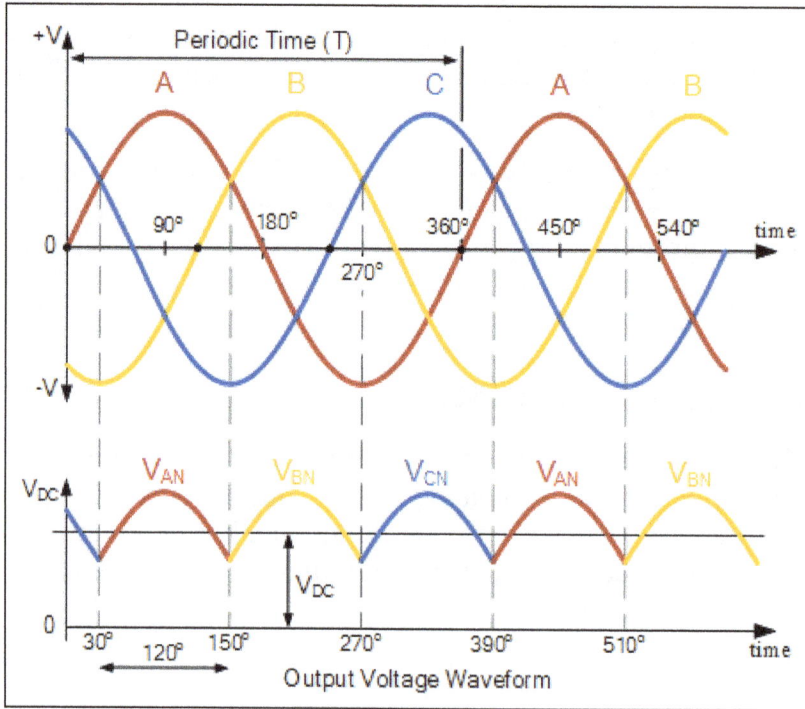

Output Voltage Waveform

From the above waveforms for a resistive load, we can see that for a half-wave rectifier each diode passes current for one third of each cycle, with the output waveform being three times the input frequency of the AC supply. Therefore there are three voltage peaks in a given cycle, so by increasing the number of phases from a single-phase to a three-phase supply, the rectification of the supply is improved, that is the output DC voltage is smoother.

For a three-phase half-wave rectifier, the supply voltages V_A V_B and V_C are balanced but with a phase difference of 120° giving:

$$V_A = V_p*\sin(\omega t - 0°),$$

$$V_B = V_p*\sin(\omega t - 120°),$$

$$V_C = V_p*\sin(\omega t - 240°).$$

Thus the average DC value of the output voltage waveform from a 3-phase half-wave rectifier is given as:

$$V_{DC} = \frac{3\sqrt{3}}{2\pi}V_p = 0.827*V_{PEAK}$$

As the voltage supplies peak voltage, V_p is equal to $V_{RMS}*1.414$, it follows that V_p is equal to $V_p/1.414$ giving $0.707*V_p$, so the average DC output voltage of the rectifier can be expressed in terms of the rms (root-mean-squared) phase voltage giving:

$$V_{DC} = \frac{3\sqrt{3}}{2\pi} \times \frac{V_{PEAK}}{1.414}$$

$$V_{DC} = \frac{0.827}{0.707} V_{RMS} = 1.17 * V_{RMS}$$

3-phase Rectification

A half-wave 3-phase rectifier is constructed using three individual diodes and a 120VAC 3-phase star connected transformer. If it is required to power a connected load with an impedance of 50Ω, Calculate, a) the average DC voltage output to the load. b) the load current, c) the average current per diode. Assume ideal diodes.

- The average DC load voltage:

  V_{DC} = 1.17*Vrms = 1.17*120 = 140.4 volts.

  If we were given the peak voltage (V_p) value,

  then:

  V_{DC} would equal 0.827*Vp or 0.827*169.68 = 140.4V.

- The DC load current:

  $I_L = V_{DC}/R_L$ = 140.4/50 = 2.81 amperes.

- The average current per diode:

  $I_D = I_L/3$ = 2.81/3 = 0.94 amperes.

One of the disadvantages of half-wave 3-phase rectification is that it requires a 4-wire supply, that is three phases plus a neutral (N) connection. Also the average DC output voltage is low at a value represented by $0.827*V_p$ as we have seen. This is because the output ripple content is three times the input frequency. But we can improve on these disadvantages by adding three more diodes to the basic rectifier circuit creating a three-phase full-wave uncontrolled bridge rectifier.

Full-wave Three-phase Rectification

The full-wave three-phase uncontrolled bridge rectifier circuit uses six diodes, two per phase in a similar fashion to the single-phase bridge rectifier. A 3-phase full-wave rectifier is obtained by using two half-wave rectifier circuits. The advantage here is that the circuit produces a lower ripple output than the previous half-wave 3-phase rectifier as it has a frequency of six times the input AC waveform.

Also, the full-wave rectifier can be fed from a balanced 3–phase 3-wire delta connected supply as no fourth neutral (N) wire is required. Consider the full-wave 3-phase rectifier circuit below.

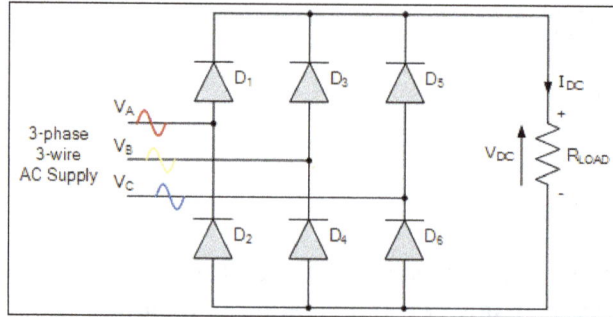

As before, assuming a phase rotation of Red-Yellow-Blue ($V_A - V_B - V_C$) and the red phase (V_A) starts at 0°. Each phase connects between a pair of diodes as shown. One diode of the conducting pair powers the positive (+) side of load, while the other diode powers the negative (-) side of load.

Diodes D_1 D_3 D_2 and D_4 form a bridge rectifier network between phases A and B, similarly diodes D_3 D_5 D_4 and D_6 between phases B and C and D_5 D_1 D_6 and D_2 between phases C and A.

Thus diodes D_1 D_3 and D_5 feed the positive rail and depending on which one has a more positive voltage at its anode terminal conducts. Likewise, diodes D_2 D_4 and D_6 feed the negative rail and whichever diode has a more negative voltage at its cathode terminal conducts.

Then we can see that for three-phase rectification, the diodes conduct in matching pairs giving a conduction pattern for the load current of: D_{1-2} D_{1-6} D_{3-6} D_{3-6} D_{3-4} D_{5-4} D_{5-2} and D_{1-2} as shown.

Full-wave Three-phase Rectifier Conduction Waveform

In 3-phase power rectifiers, conduction always occurs in the most positive diode and the corresponding most negative diode. Thus as the three phases rotate across the rectifier terminals, conduction is passed from diode to diode. Then each diode conducts for 120°(one-third) in each supply cycle but as it takes two diodes to conduct in pairs, each pair of diodes will conduct for only 60° (one-sixth) of a cycle at any one time as shown.

Therefore we can correctly say that for a 3-phase rectifier being fed by "3" transformer secondaries, each phase will be separated by 360°/3 thus requiring 2*3 diodes. Note also that unlike the previous half-wave rectifier, there is no common connection between the rectifiers input and output terminals. Therefore it can be fed by a star connected or a delta connected transformer supply.

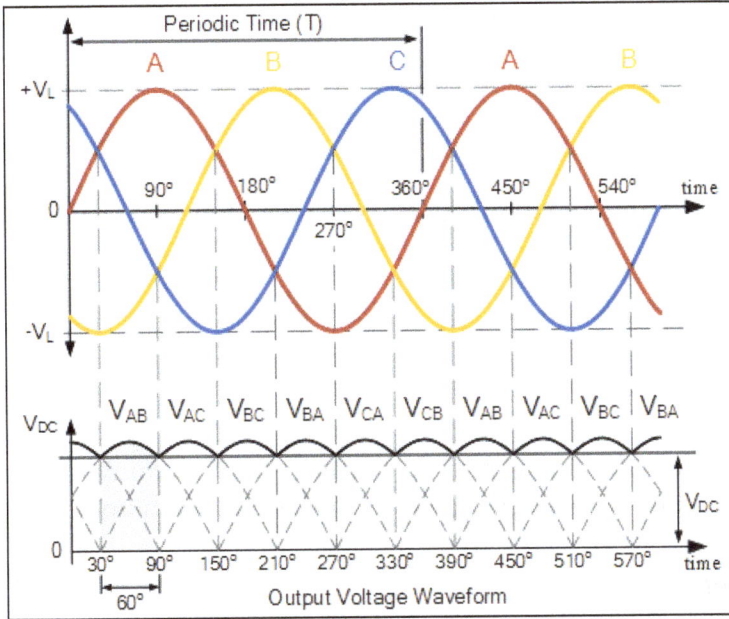

Output Voltage Waveform

So the average DC value of the output voltage waveform from a 3-phase full-wave rectifier is given as:

$$V_{DC} = \frac{3\sqrt{3}}{\pi} V_S = 1.65 * V_S$$

Where, V_S is equal to $(V_{L(PEAK)} \div \sqrt{3})$ and where $V_{L(PEAK)}$ is the maximum line-to-line voltage $(V_L * 1.414)$.

3-phase Rectification

A 3-phase full-wave bridge rectifier is required to fed a 150Ω resistive load from a 3-phase 127 volt, 60Hz delta connected supply. Ignoring the voltage drops across the diodes, calculate: 1. the DC output voltage of the rectifier and 2. the load current.

1. The DC output voltage

The RMS (Root Mean Squared) line voltage is 127 volts. Therefore the line-to-line peak voltage $(V_{L-L(PEAK)})$ will be:

$$V_{L(PEAK)} = V_{L(RMS)} \times \sqrt{2} = 127 \times 1.414 = 179.6V$$

As the supply is 3-phase, the phase to neutral voltage (V_{P-N}) of any phase will be:

$$V_S = V_{L(PEAK)} \div \sqrt{3} = 179.6 \div 1.732 = 103.7V$$

Note that this is basically the same as saying:

$$V_S = \frac{V_{L(RMS)} \times \sqrt{2}}{\sqrt{3}} = 103.7V$$

Thus the average DC output voltage from the 3-phase full-wave rectifier is given as:

$$V_{DC} = \left[\frac{3\sqrt{3}}{\pi}\right]V_S = 1.654 \times V_S$$

$$\therefore V_{DC} = 1.654 \times 103.7 = 171.5V$$

Again, we can reduce the maths a bit by correctly saying that for a given line-to-line RMS voltage value, in our example 127 volts, the average DC output voltage is:

$$V_{DC} = \frac{3\sqrt{2}}{\pi}V_{L(RMS)} = 1.35 \times 127 = 171.5V$$

2. The rectifiers load current

The output from the rectifier is feeding a 150Ω resistive load. Then using Ohms law the load current will be:

$$I_{LOAD} = V_S \div R_L = 171.5 \div 150 = 1.14 \; Amps.$$

Uncontrolled 3-phase rectification uses diodes to provide an average output voltage of a fixed value relative to the value of the input AC voltages. But to vary the output voltage of the rectifier we need to replace the uncontrolled diodes, either some or all of them, with thyristors to create what are called half-controlled or fully-controlled bridge rectifiers.

Thyristors are three terminal semiconductor devices and when a suitable trigger pulse is applied to the the thyristors gate terminal when its Anode–to–Cathode terminal voltage is positive, the device will conduct and pass a load current. So by delaying the timing of the trigger pulse, (firing angle) we can delay the instant in time at which the thyristor would naturally switch "ON" if it were a normal diode and the moment it starts to conduct when the trigger pulse is applied.

Thus with a controlled 3-phase rectification which uses thyristors instead of diodes, we can control the value of the average DC output voltage by controlling the firing angle of the thyristor pairs and so the rectified output voltage becomes a function of the firing angle, α.

Therefore the only difference to the formula used above for the average output voltage of a 3-phase bridge rectifier is in the cosine angle, cos(α) of the firing or triggering pulse. So if the firing angle is zero, (cos(0) = 1), the controlled rectifier performs similar to the previous 3-phase uncontrolled diode rectifier with the average output voltages being the same.

An example of a fully-controlled 3-phase bridge rectifier is given below:

Fully-controlled 3-phase Bridge Rectifier

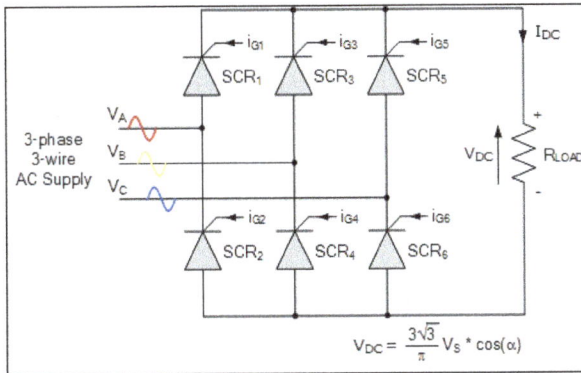

$$V_{DC} = \frac{3\sqrt{3}}{\pi} V_S * \cos(\alpha)$$

References

- Rectification: miniphysics.com, Retrieved 21 July, 2019

- Rectifier-whatisrectifier, rectifier, electronic-devices-and-circuits: physics-and-radio-electronics.com, Retrieved 22 August, 2019

- Half-wave-rectifiers: electrical4u.com, Retrieved 23 January, 2019

- Single-phase-half-wave-controlled-rectifier: electricalbaba.com, Retrieved 24 February, 2019

- Single-phase-full-wave-controlled-rectifier, ac-dc-power-converters: electronics-tutorial.net, Retrieved 25 March, 2019

- 3-phase-rectifie: electronicscoach.com, Retrieved 26 April, 2019

- Three-phase-rectification, power: electronics-tutorials.ws, Retrieved 27 May, 2019

6

Inverters

A power electronic device that changes direct current into alternating current is known as an inverter. The major types of inverters are single-phase inverter and three-phase inverter. The diverse types of inverters and their applications have been thoroughly discussed in this chapter.

A power inverter is a piece of equipment that transforms direct current (DC) electricity to alternating current (AC) electricity. Batteries produce direct current electricity, and a common implementation of power inverters is car batteries as a source for indoor power during power outages (for common household appliances that are powered by alternating current). Power inverters are sold as consumer items and are typically simple to install.

The earliest power inverters were complex electromechanical devices that used magnets and moving parts (including spring arms) to transduce direct current power into alternating current electricity. Modern power inverters employ oscillator circuits, which are made up of transistors and semiconductors rather than elaborate moving parts.

SINGLE PHASE INVERTER

There are two types of single phase inverters – full bridge inverter and half bridge inverter.

Half Bridge Inverter

Single Phase Half Bridge Inverter has two thyristors and two free-wheeling diodes. Each thyristor is gated at frequency.

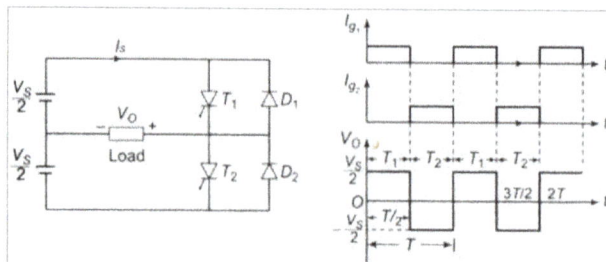

Single-phase half-bridge inverter.

$f = 1/T$ of the AC supply desired. The gating signals of the two thyristors have a phase angle of 180°. From figure, the output is easily seen to be rectangular ac waveform of frequency.

$$\omega = \frac{2\pi}{T} \; rad/s$$

where, T = triggering period of the thyristor.

The output waveform feeds the load which may in general comprise RLC components. The circuit model of the inverter is given in figure. After several cycles of source voltage vTh have elapsed, the time variation of current settles down to periodic form such that:

$$i_0 = \mp I_{01}; \qquad t = T/2$$
$$i_0 = \pm I_{01}; \qquad t = 0, T$$

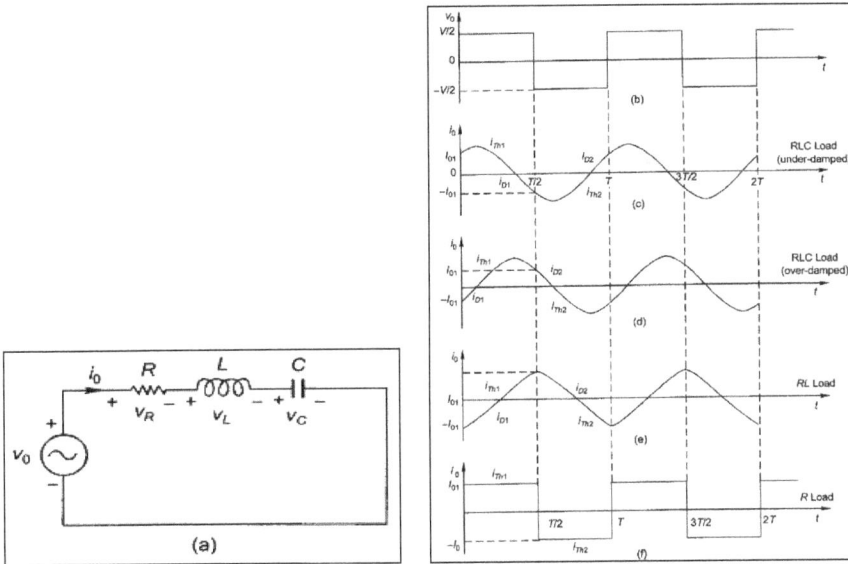

Circuit model and waveforms for the inverter.

During the interval 0 < t < T/2,

$$v_0 = \frac{V}{2} = v_R + v_L + v_C$$

$$\frac{V}{2} = Ri_0 + L\frac{di_0}{dt} + \frac{1}{C}\int_0^i i_0 \, dt + V_{C0}$$

where, V_{c0} is the voltage across the capacitive element at t = 0. Differentiating above equation,

$$\frac{d^2 i_0}{dt^2} + \frac{R}{L}\frac{di_0}{dt} + \frac{1}{LC}i_0 = 0$$

The nature of the waveform will depend upon the circuit damping. The output voltage waveform (rectangular) and various current waveforms for different load characteristics are drawn in figure. In the RLC load underdamped case of figure, the current of thyristor Th_1 becomes zero and the thyristor turns off before Th_2 is gated. The circuit conditions cause the diode D_1 to become forward-biased conducting the free-wheeling current iD1. As soon as Th_2 is triggered, D_1 is reverse-biased and the current is transferred from D_1 to Th_2. This process then repeats for Th_2 and D_2 and so on. In the RLC load overdamped case of figure and the RL load of figure, Th_1 is forced to switch off (at $T/_2$) while it is still carrying current. As a consequence, the voltage across the L-component of the load reverses, causing D_2 to become forward-biased which then conducts the free-wheeling current. As the circuit current tends to reverse, D_2 switches off and the current is conducted by Th_2 which has already been gated at $t = T/_2$. For the ideal R load case of figure, there is theoretically no free-wheeling current. This does not mean that the free-wheeling diode can be done away with as any practical circuit always has some inductance.

Full Bridge Inverter

A single phase bridge DC-AC inverter is shown in figure below. The analysis of the single phase DC-AC inverters is done taking into account following assumptions and conventions.

1. The current entering node a in figure is considered to be positive.

2. The switches S1, S2, S3 and S4 are unidirectional, i.e. they conduct current in one direction.

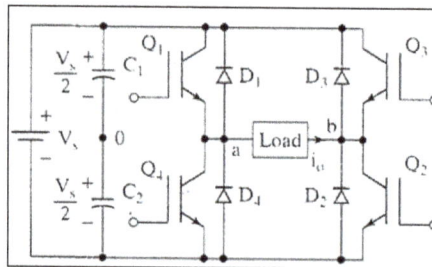

When the switches S1 and S2 are turned on simultaneously for a duration $0 \leq t \leq T_1$, the the input voltage Vin appears across the load and the current flows from point a to b.

Q1 – Q2 ON, Q3 – Q4 OFF ==> v o = Vs

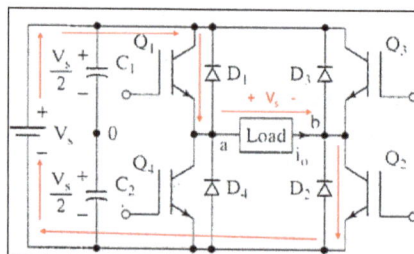

If the switches S3 and S4 turned on duration T1 ≤ t ≤ T2, the voltage across the load the load is reversed and the current through the load flows from point b to a.

Q1 – Q2 OFF, Q3 – Q4 ON ==> v o = -Vs.

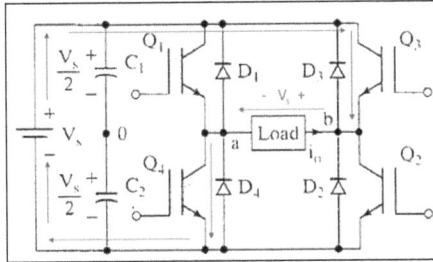

The voltage and current waveforms across the resistive load are shown in figure below:

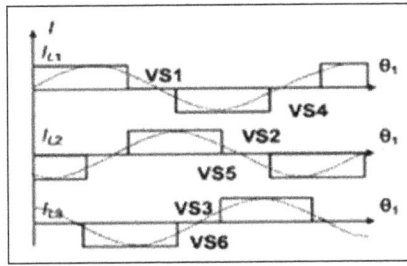

Single Phase Full Bridge Inverter for R-L load

A single-phase square wave type voltage source inverter produces square shaped output voltage for a single-phase load. Such inverters have very simple control logic and the power switches need to operate at much lower frequencies compared to switches in some other types of inverters. The first generation inverters, using thyristor switches, were almost invariably square wave inverters because thyristor switches could be switched on and off only a few hundred times in a second. In contrast, the present day switches like IGBTs are much faster and used at switching frequencies of several kilohertz. Single-phase inverters mostly use half bridge or full bridge topologies. Power circuits of these topologies are shown in figure below.

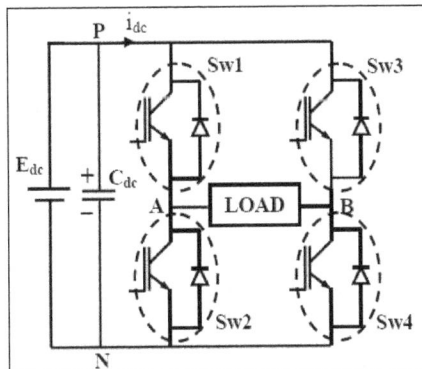

The above topology are analyzed under the assumption of ideal circuit conditions. Accordingly, it is assumed that the input DC voltage (Edc) is constant and the switches are lossless. In full bridge topology has two such legs. Each leg of the inverter consists of two series connected electronic switches shown within dotted lines in the figures. Each of these switches consists of an IGBT type controlled switch across which an uncontrolled diode is put in anti-parallel manner. These switches are capable of conducting bi-directional current but they need to block only one polarity of voltage. The junction point of the switches in each leg of the inverter serves as one output point for the load.

THREE PHASE INVERTER

A three-phase inverter converts a DC input into a three-phase AC output. Its three arms are normally delayed by an angle of 120° so as to generate a three-phase AC supply. The inverter switches each has a ratio of 50% and switching occurs after every T/6 of the time T (60° angle interval). The switches S1 and S4, the switches S2 and S5 and switches S3 and S6 complement each other.

The figure below shows a circuit for a three phase inverter. It is nothing but three single phase inverters put across the same DC source. The pole voltages in a three phase inverter are equal to the pole voltages in single phase half bridge inverter.

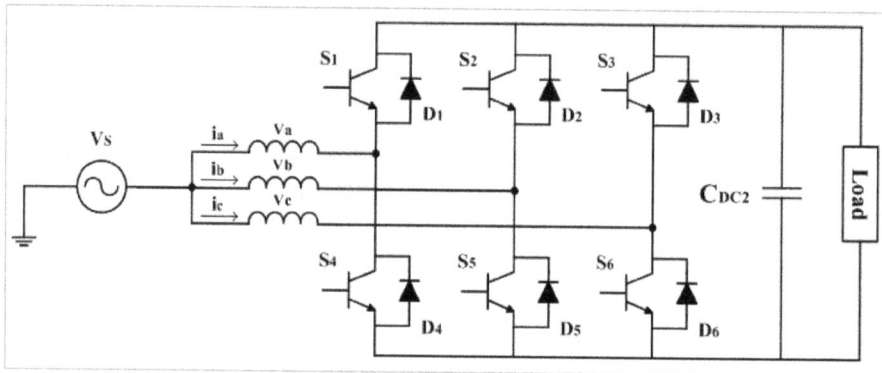

The two types of inverters above have two modes of conduction – 180° mode of conduction and 120° mode of conduction.

1. 180° mode of conduction: In this mode of conduction, every device is in conduction state for 180° where they are switched ON at 60° intervals. The terminals A, B and C are the output terminals of the bridge that are connected to the three-phase delta or star connection of the load.

The operation of a balanced star connected load is explained in the diagram below. For the period 0° – 60° the points S1, S5 and S6 are in conduction mode. The terminals A and C of the load are connected to the source at its positive point. The terminal B is connected to the source at its negative point. In addition, resistances R/2 is between the neutral and the positive end while resistance R is between the neutral and the negative terminal.

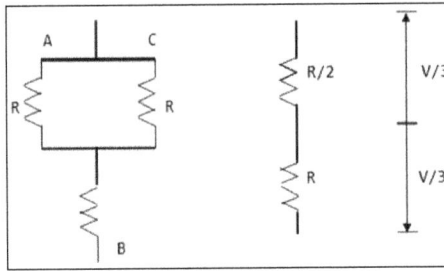

The load voltages are gives as follows:

$$V_{AN} = V/3,$$

$$V_{BN} = -2V/3,$$

$$V_{CN} = V/3.$$

The line voltages are given as follows:

$$V_{AB} = V_{AN} - V_{BN} = V,$$

$$V_{BC} = V_{BN} - V_{CN} = -V,$$

$$V_{CA} = V_{CN} - V_{AN} = 0.$$

Waveforms for 180° mode of conduction.

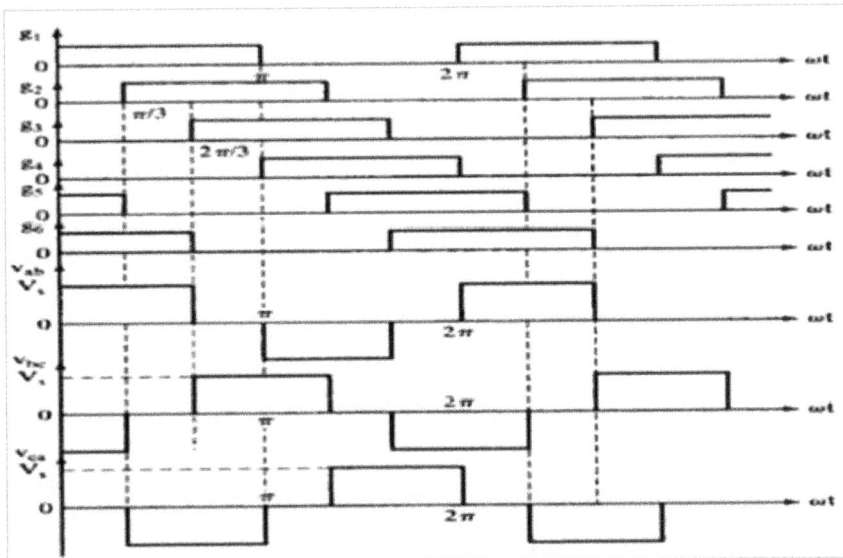

2. 120° mode of conduction: In this mode of conduction, each electronic device is in a conduction state for 120°. It is most suitable for a delta connection in a load because it

results in a six-step type of waveform across any of its phases. Therefore, at any instant only two devices are conducting because each device conducts at only 120°.

The terminal A on the load is connected to the positive end while the terminal B is connected to the negative end of the source. The terminal C on the load is in a condition called floating state. Furthermore, the phase voltages are equal to the load voltages as shown below:

Phase voltages = Line voltages.

$$V_{AB} = V,$$

$$V_{BC} = -V/2,$$

$$V_{CA} = -V/2.$$

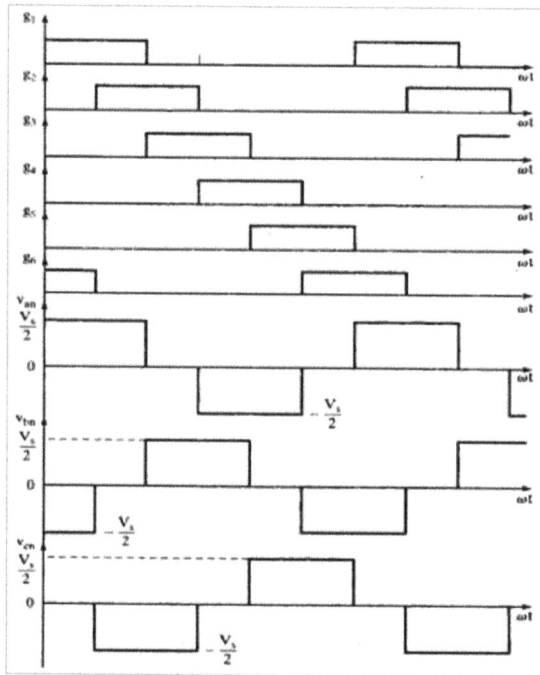

Waveforms for 120° mode of conduction.

APPLICATIONS OF INVERTERS

An inverter takes current from a battery, which is usually 12 Volt Dc, as batteries usually produce DC power, and then after passing this current through a 50Hz processor, it converts it into the normal 220 volt AC which is commonly used.

In case we want to power a small appliance like a mobile charger, an LED or some small radio, then that 12 volt DC power would be sufficient enough. But if the question is of a bigger appliance, like a fridge, TV, Laptop or something like that, then we surely have to make use of AC power since most of our bigger appliances are designed to run on AC.

In such cases, inverters are of great use, especially if the appliance is not powered directly from a switch board or is acquiring power from a battery or some sort of DC producing devices.

The process of conversion is shown through a block diagram as follows:

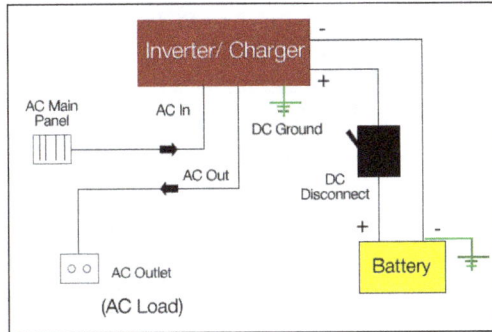

Use in Solar Panels

Other than these, bigger inverters are also available which can run on 24 volt DC as well and deal with higher currents. Other than batteries, we know that solar power is becoming the new trend now a days, since it is cheaper, economical and hence more affordable.

So it is needless to mention here that inverters find their diverse uses in solar panels as well, as the solar panels produce DC, which then needs to be converted to AC by the help of inverters.

This thing can be shown below:

Use in Fuel Cells and UPS

Another very common use of inverter is in fuel cells, as they also produce DC power which later needs to be converted to AC for use in electronic equipment.

Similarly, we have seen that UPS (Uninterruptible Power Supplies) are a very common trend now a days due to load shedding and a need of an uninterruptible power to sensitive equipment for domestic or commercial uses.

A UPS uses a set of batteries along with an inverter to supply power when the main power is no longer available.

These were some of the most common examples, in which inverters find their extensive uses in everyday life. Much more of it is still there through which we can estimate the necessity of using inverters in more than half of the electronic equipment.

So according to the need and demand, advancement in the inverter technology is still on its way and many types of inverters are being manufactured and designed according to the need.

References

- Power-inverter, definition: techopedia.com, Retrieved 28 June, 2019
- Single-phase-half-bridge-inverter: eeeguide.com, Retrieved 29 July, 2019
- Single-phase-full-bridge-inverter, c-to-ac-inverter: electronics-tutorial.net, Retrieved 30 August, 2019
- Power-electronics-types-of-inverters, power-electronics: tutorialspoint.com, Retrieved 1 January, 2019
- Applications-of-inverters, electrical-distribution: electrical-equipment.org, Retrieved 2 February, 2019

7

Converters

Power electronics converters are used to modify the form of electrical energy. Two major types of converters are DC to DC converter and AC to AC converter. This chapter discusses in detail the various subtypes of these types of converters.

The primary task of power electronics is to process and control the flow of electric energy by supplying voltages and currents in a form that is optimally suited for user loads. Modern power electronic converters are involved in a very broad spectrum of applications like switched-mode power supplies, active power filters, electrical-machine-motion-control, renewable energy conversion systems distributed power generation, flexible AC transmission systems, and vehicular technology, etc.

Power electronic converters can be found wherever there is a need to modify the electrical energy form with classical electronics in which electrical currents and voltage are used to carry information, whereas with power electronics, they carry power. Some examples of uses for power electronic systems are DC/DC converters used in many mobile devices, such as cell phones or PDAs, and AC/DC converters in computers and televisions. Large scale power electronics are used to control hundreds of megawatt of power flow across our nation.

Dual Converter

Dual converter is a combination of a rectifier and inverter in which the conversion of A.C to D.C happens and followed by D.C to A.C where load lies in between. A dual converter can be of a single phase or a three phase. A dual converter consists of two bridges consisting of thyristors in which one for rectifying purpose where alternating current is converted to direct current which can be given to load. Other bridge of thyristors is used for converting D.C to A.C.

Single Phase Dual Converter

Single phase dual converter uses a single phase as source which is given to converter 1 of dual converter for rectification followed to load.

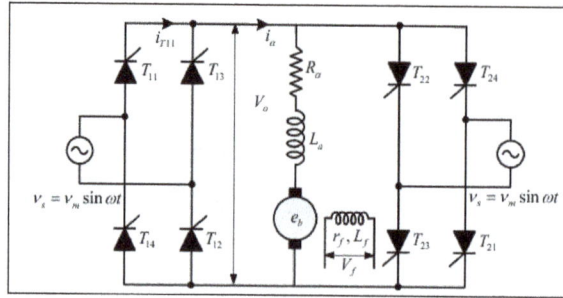

Principle of Operation

A.C input given to converter 1 for rectification in this process positive cycle of input is given to first set of forward biased thyristors which gives a rectified D.C on positive cycle, as well negative cycle is given to set of reverse biased thyristors which gives a D.C on negative cycle completing full wave rectified output can be given to load. During this process converter 2 is blocked using an inductor. As thyristor only start conducting when current pulse is given to gate and continuous conducting until supply of current is stopped. Output of Thyristor Bridge can be as follows when it is given to different loads.

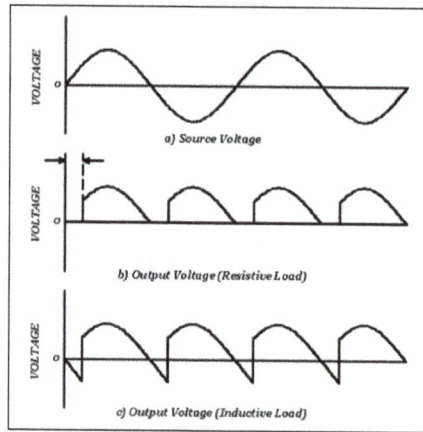

a) Source Voltage

b) Output Voltage (Resistive Load)

c) Output Voltage (Inductive Load)

As a dual converter also consists conversion of D.C to A.C to make it work converter two is blocked, D.C inputs become load to dc power source conversion.

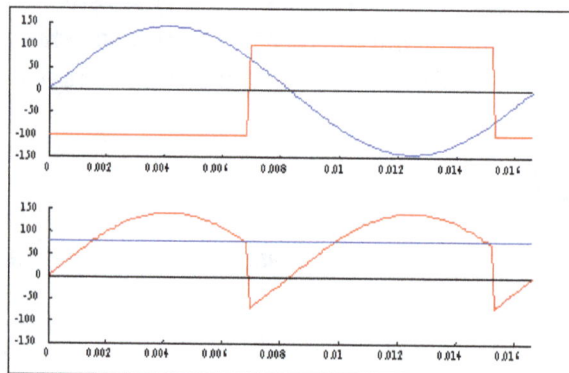

Firing of Thyristors

To make thyristors conduct, a trigger pulse must be given to its gate simultaneously along with line voltage. A separate gate drive circuit must be added to a dual converter thyristor bridges Gate drive circuit must be equally synchronized with source voltage, any delay causes zero cross jitter and zero frequency fluctuates. To prevent these circuits must be included with phase lock loops and comparators.

Applications of Single Phase Dual Converter

- Speed control and direction control in dc motors.

Speed control and Polarity Control of DC Motor using Single Phase Dual Converter

A single phase dual converter can be used in controlling speed and direction of rotation interfacing with microcontroller, combination of four SCR's is placed either side of motor and motor is load. These thyristors can be triggered through an optocoupler which is connected to a port of microcontroller.

Rotation of motor can be initialized using optocoupler by setting a set of thyristor to trigger which is placed at one side and change in direction of motor can be achieved by triggering another set of thyristor Variation in speed of motor can be achieved by delayed firing angle of SCR.

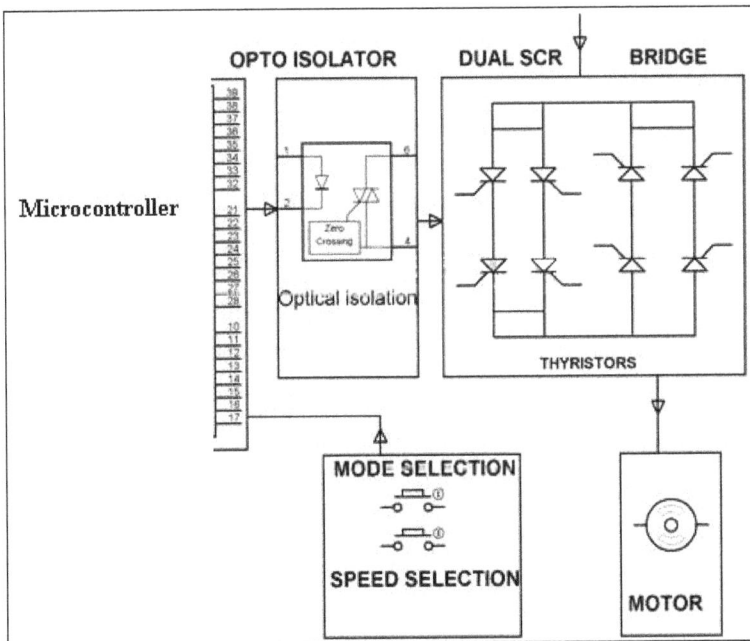

Mode selection and speed selection are microcontroller interfaced switches using these switches speed and rotation can be selected.

DC TO DC CONVERTER

A DC-to-DC converter is an electronic circuit or electromechanical device that converts a source of direct current (DC) from one voltage level to another. It is a type of electric power converter. Power levels range from very low (small batteries) to very high (high-voltage power transmission).

Uses

DC to DC converters are used in portable electronic devices such as cellular phones and laptop computers, which are supplied with power from batteries primarily. Such electronic devices often contain several sub-circuits, each with its own voltage level requirement different from that supplied by the battery or an external supply (sometimes higher or lower than the supply voltage). Additionally, the battery voltage declines as its stored energy is drained. Switched DC to DC converters offer a method to increase voltage from a partially lowered battery voltage thereby saving space instead of using multiple batteries to accomplish the same thing.

Most DC to DC converter circuits also regulate the output voltage. Some exceptions include high-efficiency LED power sources, which are a kind of DC to DC converter that regulates the current through the LEDs, and simple charge pumps which double or triple the output voltage.

DC to DC converters which are developed to maximize the energy harvest for photovoltaic systems and for wind turbines are called power optimizers.

Transformers used for voltage conversion at mains frequencies of 50–60 Hz must be large and heavy for powers exceeding a few watts. This makes them expensive, and they are subject to energy losses in their windings and due to eddy currents in their cores. DC-to-DC techniques that use transformers or inductors work at much higher frequencies, requiring only much smaller, lighter, and cheaper wound components. Consequently these techniques are used even where a mains transformer could be used; for example, for domestic electronic appliances it is preferable to rectify mains voltage to DC, use switch-mode techniques to convert it to high-frequency AC at the desired voltage, then, usually, rectify to DC. The entire complex circuit is cheaper and more efficient than a simple mains transformer circuit of the same output.

Electronic Conversion

Practical electronic converters use switching techniques. Switched-mode DC-to-DC converters convert one DC voltage level to another, which may be higher or lower, by storing the input energy temporarily and then releasing that energy to the output at a different voltage. The storage may be in either magnetic field storage components (inductors, transformers) or electric field storage components (capacitors). This conversion method can increase or decrease voltage. Switching conversion is often more power-efficient (typical efficiency is 75% to 98%) than linear voltage regulation, which dissipates unwanted power as heat. Fast semiconductor device rise and fall times are required for efficiency; however, these fast transitions combine with layout parasitic effects to make circuit design challenging. The higher efficiency of a switched-mode converter reduces the heatsinking needed, and increases battery endurance of portable equipment. Efficiency has improved since the late 1980s due to the use of power FETs, which are able to switch more efficiently with lower switching losses at higher frequencies than power bipolar transistors, and use less complex drive circuitry. Another important improvement in DC-DC converters is replacing the flywheel diode by synchronous rectification using a power FET, whose "on resistance" is much lower, reducing switching losses. Before the wide availability of power semiconductors, low-power DC-to-DC synchronous converters consisted of an electro-mechanical vibrator followed by a voltage step-up transformer feeding a vacuum tube or semiconductor rectifier, or synchronous rectifier contacts on the vibrator.

Most DC-to-DC converters are designed to move power in only one direction, from dedicated input to output. However, all switching regulator topologies can be made bidirectional and able to move power in either direction by replacing all diodes with independently controlled active rectification. A bidirectional converter is useful, for example, in applications requiring regenerative braking of vehicles, where power is supplied *to* the wheels while driving, but supplied *by* the wheels when braking.

Although they require few components, switching converters are electronically complex. Like all high-frequency circuits, their components must be carefully specified and physically arranged to achieve stable operation and to keep switching noise (EMI/RFI) at acceptable levels. Their cost is higher than linear regulators in voltage-dropping applications, but their cost has been decreasing with advances in chip design.

DC-to-DC converters are available as integrated circuits (ICs) requiring few additional components. Converters are also available as complete hybrid circuit modules, ready for use within an electronic assembly.

Linear regulators which are used to output a stable DC independent of input voltage and output load from a higher but less stable input by dissipating excess volt-amperes as heat, could be described literally as DC-to-DC converters, but this is not usual usage. (The same could be said of a simple voltage dropper resistor, whether or not stabilised by a following voltage regulator or Zener diode).

There are also simple capacitive voltage doubler and Dickson multiplier circuits using diodes and capacitors to multiply a DC voltage by an integer value, typically delivering only a small current.

Magnetic

In these DC-to-DC converters, energy is periodically stored within and released from a magnetic field in an inductor or a transformer, typically within a frequency range of 300 kHz to 10 MHz. By adjusting the duty cycle of the charging voltage (that is, the ratio of the on/off times), the amount of power transferred to a load can be more easily controlled, though this control can also be applied to the input current, the output current, or to maintain constant power. Transformer-based converters may provide isolation between input and output. In general, the term *DC-to-DC converter* refers to one of these switching converters. These circuits are the heart of a switched-mode power supply. Many topologies exist. This table shows the most common ones.

	Forward (energy transfers through the magnetic field)	Flyback (energy is stored in the magnetic field)
No transformer (non-isolated)	• Step-down (buck) - The output voltage is lower than the input voltage, and of the same polarity.	• Non-inverting: The output voltage is the same polarity as the input. ◦ Step-up (boost) - The output voltage is higher than the input voltage. ◦ SEPIC - The output voltage can be lower or higher than the input. • Inverting: the output voltage is of the opposite polarity as the input. ◦ Inverting (buck-boost). ◦ Ćuk - Output current is continuous.
	• True buck-boost - The output voltage is the same polarity as the input and can be lower or higher.	
	• Split-pi (boost-buck) - Allows bidirectional voltage conversion with the output voltage the same polarity as the input and can be lower or higher.	
With transformer (isolatable)	• Forward - 1 or 2 transistor drive. • Push-pull (half bridge) - 2 transistors drive. • Full bridge - 4 transistor drive.	• Flyback - 1 transistor drive.

In addition, each topology may be:

• Hard Switched: Transistors switch quickly while exposed to both full voltage and full current.

• Resonant: An LC circuit shapes the voltage across the transistor and current through it so that the transistor switches when either the voltage or the current is zero.

Magnetic DC-to-DC converters may be operated in two modes, according to the current in its main magnetic component (inductor or transformer):

• Continuous: The current fluctuates but never goes down to zero.

• Discontinuous: The current fluctuates during the cycle, going down to zero at or before the end of each cycle.

A converter may be designed to operate in continuous mode at high power, and in discontinuous mode at low power.

The half bridge and flyback topologies are similar in that energy stored in the magnetic core needs to be dissipated so that the core does not saturate. Power transmission in a flyback circuit is limited by the amount of energy that can be stored in the core, while forward circuits are usually limited by the I/V characteristics of the switches.

Although MOSFET switches can tolerate simultaneous full current and voltage (although thermal stress and electromigration can shorten the MTBF), bipolar switches generally can't so require the use of a snubber (or two).

High-current systems often use multiphase converters, also called interleaved converters. Multiphase regulators can have better ripple and better response times than single-phase regulators.

Many laptop and desktop motherboards include interleaved buck regulators, sometimes as a voltage regulator module.

Capacitive

Switched capacitor converters rely on alternately connecting capacitors to the input and output in differing topologies. For example, a switched-capacitor reducing converter might charge two capacitors in series and then discharge them in parallel. This would produce the same output power (less that lost to efficiency of under 100%) at, ideally, half the input voltage and twice the current. Because they operate on discrete quantities of charge, these are also sometimes referred to as charge pump converters. They are typically used in applications requiring relatively small currents, as at higher currents the increased efficiency and smaller size of switch-mode converters makes them a better choice. They are also used at extremely high voltages, as magnetics would break down at such voltages.

Electromechanical Conversion

A motor generator with separate motor and generator.

A motor-generator set, mainly of historical interest, consists of an electric motor and generator coupled together. A *dynamotor* combines both functions into a single unit with coils for both the motor and the generator functions wound around a single rotor; both coils share the same outer field coils or magnets. Typically the motor coils are driven from a commutator on one end of the shaft, when the generator coils output to another commutator on the other end of the shaft. The entire rotor and shaft assembly is smaller in size than a pair of machines, and may not have any exposed drive shafts.

Motor-generators can convert between any combination of DC and AC voltage and phase standards. Large motor-generator sets were widely used to convert industrial amounts of power while smaller units were used to convert battery power (6, 12 or 24 V DC) to a high DC voltage, which was required to operate vacuum tube (thermionic valve) equipment.

For lower-power requirements at voltages higher than supplied by a vehicle battery, vibrator or "buzzer" power supplies were used. The vibrator oscillated mechanically, with contacts that switched the polarity of the battery many times per second, effectively converting DC to square wave AC, which could then be fed to a transformer of the required output voltage(s). It made a characteristic buzzing noise.

Electrochemical Conversion

A further means of DC to DC conversion in the kilowatts to megawatts range is presented by using redox flow batteries such as the vanadium redox battery.

Chaotic Behavior

DC-to-DC converters are subject to different types of chaotic dynamics such as bifurcation, crisis, and intermittency.

Step-down

A converter where output voltage is lower than the input voltage (such as a buck converter).

Step-up

A converter that outputs a voltage higher than the input voltage (such as a boost converter).

Continuous Current Mode

Current and thus the magnetic field in the inductive energy storage never reaches zero.

Discontinuous Current Mode

Current and thus the magnetic field in the inductive energy storage may reach or cross zero.

Noise

Unwanted electrical and electromagnetic signal noise, typically switching artifacts.

RF Noise

Switching converters inherently emit radio waves at the switching frequency and its harmonics. Switching converters that produce triangular switching current, such as the Split-Pi, forward converter, or Ćuk converter in continuous current mode, produce less harmonic noise than other switching converters. RF noise causes electromagnetic interference (EMI). Acceptable levels depend upon requirements, e.g. proximity to RF circuitry needs more suppression than simply meeting regulations.

Input Noise

The input voltage may have non-negligible noise. Additionally, if the converter loads the input with sharp load edges, the converter can emit RF noise from the supplying power lines. This should be prevented with proper filtering in the input stage of the converter.

Output Noise

The output of an ideal DC-to-DC converter is a flat, constant output voltage. However, real converters produce a DC output upon which is superimposed some level of electrical noise. Switching converters produce switching noise at the switching frequency and its harmonics. Additionally, all electronic circuits have some thermal noise. Some sensitive radio-frequency and analog circuits require a power supply with so little noise that it can only be provided by a linear regulator. Some analog circuits which require a power supply with relatively low noise can tolerate some of the less-noisy switching converters, e.g. using continuous triangular waveforms rather than square waves.

Buck Converter

A buck converter (step-down converter) is a DC-to-DC power converter which steps down voltage (while stepping up current) from its input (supply) to its output (load). It is a class of switched-mode power supply (SMPS) typically containing at least two semiconductors (a diode and a transistor, although modern buck converters frequently replace the diode with a second transistor used for synchronous rectification) and at least one energy storage element, a capacitor, inductor, or the two in combination. To reduce voltage ripple, filters made of capacitors (sometimes in combination with inductors) are normally added to such a converter's output (load-side filter) and input (supply-side filter).

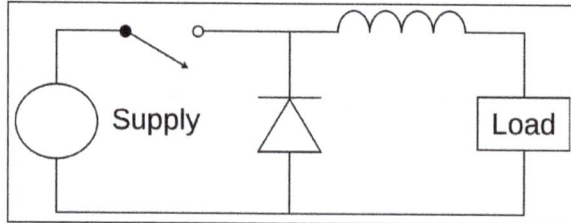

Switching converters (such as buck converters) provide much greater power efficiency as DC-to-DC converters than linear regulators, which are simpler circuits that lower voltages by dissipating power as heat, but do not step up output current.

Two small commodity buck converter modules and one boost.

Buck converters can be highly efficient (often higher than 90%), making them useful for tasks such as converting a computer's main (bulk) supply voltage (often 12 V) down to lower voltages needed by USB, DRAM and the CPU (1.8 V or less).

Theory of Operation

The basic operation of the buck converter has the current in an inductor controlled by two switches (usually a transistor and a diode). In the idealised converter, all the components are considered to be perfect. Specifically, the switch and the diode have zero voltage drop when on and zero current flow when off, and the inductor has zero series resistance. Further, it is assumed that the input and output voltages do not change over the course of a cycle (this would imply the output capacitance as being infinite).

The conceptual model of the buck converter is best understood in terms of the relation between current and voltage of the inductor. Beginning with the switch open (off-state), the current in the circuit is zero. When the switch is first closed (on-state), the current will begin to increase, and the inductor will produce an opposing voltage across its terminals in response to the changing current. This voltage drop counteracts the voltage of the source and therefore reduces the net voltage across the load. Over time, the rate of change of current decreases, and the voltage across the inductor also then decreases, increasing the voltage at the load. During this time, the inductor stores energy in the form of a magnetic field. If the switch is opened while the current is still changing, then there will always be a voltage drop across the inductor, so the net voltage at the load will always be less than the input voltage source. When the switch is opened again (off-state), the voltage source will be removed from the circuit, and the current will decrease. The decreasing current will produce a voltage drop across the inductor (opposite to the drop at on-state), and now the inductor becomes a Current Source. The stored energy in the inductor's magnetic field supports the current flow through the load. This current, flowing while the input voltage source is disconnected, when concatenated with the current flowing during on-state, totals to current greater than the average input current (being zero during off-state). The "increase" in average current makes up for the reduction in voltage, and ideally preserves the power provided to the load. During the off-state, the inductor is discharging its stored energy into the rest of the circuit. If the switch is closed again before the inductor fully discharges (on-state), the voltage at the load will always be greater than zero.

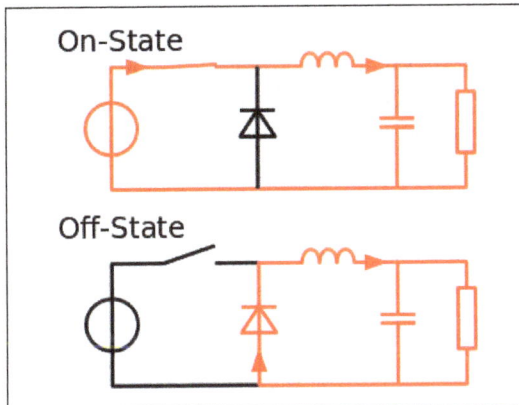

The two circuit configurations of a buck converter: on-state, when the switch is closed; and off-state, when the switch is open.

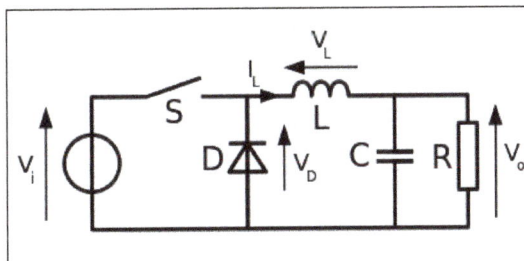

Naming conventions of the components, voltages and current of the buck converter.

Continuous Mode

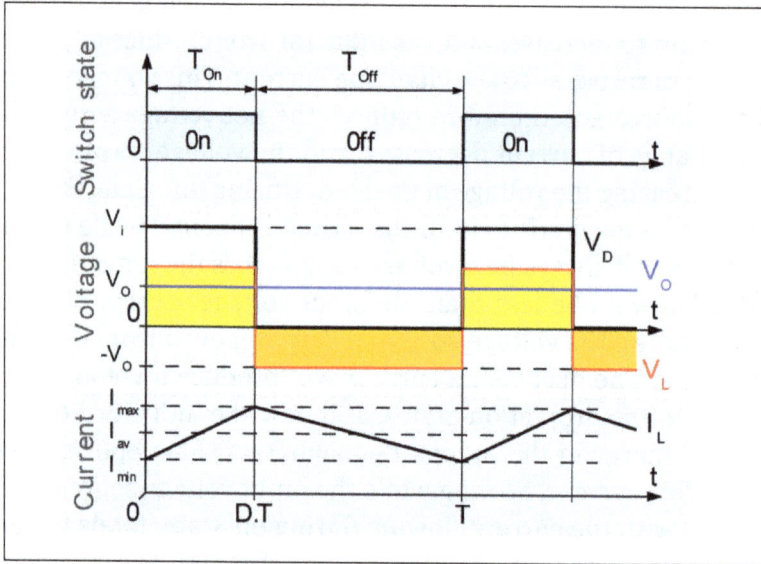

Evolution of the voltages and currents with time in an
ideal buck converter operating in continuous mode.

Buck converter operates in continuous mode if the current through the inductor (I_L)
never falls to zero during the commutation cycle. In this mode, the operating principle
is described by the plots:

- When the switch pictured above is closed, the voltage across the inductor is
  $V_L = V_i - V_o$. The current through the inductor rises linearly (in approximation,
  so long as the voltage drop is almost constant). As the diode is reverse-biased by
  the voltage source V, no current flows through it.

- When the switch is opened, the diode is forward biased. The voltage across the
  inductor is $V_L = -V_o$ (neglecting diode drop). Current I_L decreases.

The energy stored in inductor L is:

$$E = \frac{1}{2} L I_L^2$$

Therefore, it can be seen that the energy stored in L increases during on-time as I_L
increases and then decreases during the off-state. L is used to transfer energy from the
input to the output of the converter.

The rate of change of I_L can be calculated from:

$$V_L = L \frac{dI_L}{dt}$$

With V_L equal to $V_i - V_o$ during the on-state and to $-V_o$ during the off-state. Therefore, the increase in current during the on-state is given by:

$$\Delta I_{L_{on}} = \int_0^{t_{on}} \frac{V_L}{L}\,dt = \frac{(V_i - V_o)}{L} t_{on} , t_{on} = DT$$

where D is a scalar called the duty cycle with a value between 0 and 1.

Conversely, the decrease in current during the off-state is given by:

$$\Delta I_{L_{off}} = \int_{t_{on}}^{T = t_{on} + t_{off}} \frac{V_L}{L}\,dt = -\frac{V_o}{L} t_{off} , t_{off} = (1 - D)T$$

If we assume that the converter operates in the steady state, the energy stored in each component at the end of a commutation cycle T is equal to that at the beginning of the cycle. That means that the current I_L is the same at t = 0 and at t = T.

So we can write from the above equations:

$$\Delta I_{L_{on}} + \Delta I_{L_{off}} = 0$$
$$\frac{V_i - V_o}{L} t_{on} - \frac{V_o}{L} t_{off}$$

The above integrations can be done graphically. In figure, $\Delta I_{L_{on}}$ is proportional to the area of the yellow surface, and $\Delta I_{L_{off}}$ to the area of the orange surface, as these surfaces are defined by the inductor voltage (red lines). As these surfaces are simple rectangles, their areas can be found easily: $(V_i - V_o)t_{on}$ for the yellow rectangle and $-V_o t_{off}$ for the orange one. For steady state operation, these areas must be equal.

As can be seen in figure, $t_{on} = DT$ and $t_{off} = (1 - D)T$.

This yields:

$$(V_i - V_o)DT - V_o(1 - D)T = 0$$
$$DV_i - V_o = 0$$
$$\Rightarrow D = \frac{V_o}{V_i}$$

From this equation, it can be seen that the output voltage of the converter varies linearly with the duty cycle for a given input voltage. As the duty cycle D is equal to the ratio between t_{on} and the period T, it cannot be more than 1. Therefore, $V_o \leq V_i$. This is why this converter is referred to as step-down converter.

So, for example, stepping 12 V down to 3 V (output voltage equal to one quarter of the input voltage) would require a duty cycle of 25%, in our theoretically ideal circuit.

Discontinuous Mode

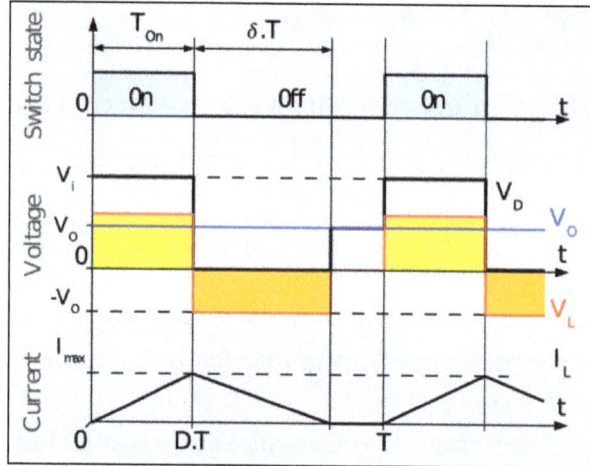

Evolution of the voltages and currents with time in an ideal buck converter operating in discontinuous mode.

In some cases, the amount of energy required by the load is too small. In this case, the current through the inductor falls to zero during part of the period. The only difference in the principle described is that the inductor is completely discharged at the end of the commutation cycle. This has, however, some effect on the previous equations.

The inductor current falling below zero results in the discharging of the output capacitor during each cycle and therefore higher switching losses. A different control technique known as Pulse-frequency modulation can be used to minimize these losses.

We still consider that the converter operates in steady state. Therefore, the energy in the inductor is the same at the beginning and at the end of the cycle (in the case of discontinuous mode, it is zero). This means that the average value of the inductor voltage (V_L) is zero; i.e., that the area of the yellow and orange rectangles in figure are the same. This yields:

$$\left(V_i - V_o\right)DT - V_o \delta T = 0$$

So the value of δ is:

$$\delta = \frac{V_i - V_o}{V_o} D$$

The output current delivered to the load (I_o) is constant, as we consider that the output capacitor is large enough to maintain a constant voltage across its terminals during a commutation cycle. This implies that the current flowing through the capacitor has a zero average value. Therefore, we have:

$$\overline{I_L} = I_o$$

Where $\overline{I_L}$ is the average value of the inductor current. As can be seen in figure, the inductor current waveform has a triangular shape. Therefore, the average value of IL can be sorted out geometrically as follow:

$$\overline{I_L} = \left(\frac{1}{2} I_{L_{max}} DT + \frac{1}{2} I_{L_{max}} \delta T \right) \frac{1}{T}$$

$$= \frac{I_{L_{max}}(D+\delta)}{2}$$

$$= I_o$$

The inductor current is zero at the beginning and rises during t_{on} up to I_{Lmax}. That means that I_{Lmax} is equal to:

$$I_{L_{max}} = \frac{V_i - V_o}{L} DT$$

Substituting the value of I_{Lmax} in the previous equation leads to:

$$I_o = \frac{(V_i - V_o) DT (D+\delta)}{2L}$$

And substituting δ by the expression given above yields:

$$I_o = \frac{(V_i - V_o) DT \left(D + \frac{V_i - V_o}{V_o} D \right)}{2L}$$

This expression can be rewritten as:

$$V_o = V_i \frac{1}{\frac{2LI_o}{D^2 V_i T} + 1}$$

It can be seen that the output voltage of a buck converter operating in discontinuous mode is much more complicated than its counterpart of the continuous mode. Furthermore, the output voltage is now a function not only of the input voltage (V_i) and the duty cycle D, but also of the inductor value (L), the commutation period (T) and the output current (I_o).

From Discontinuous to Continuous Mode and Vice Versa

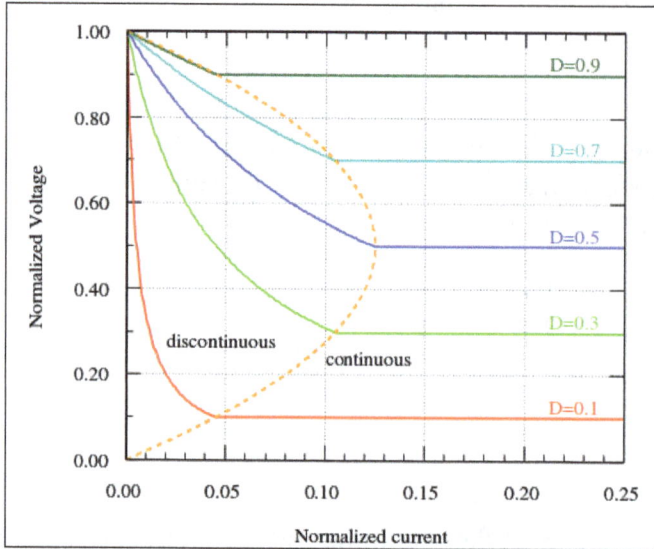

Evolution of the normalized output voltages with the normalized output current.

As mentioned at the beginning, the converter operates in discontinuous mode when low current is drawn by the load, and in continuous mode at higher load current levels. The limit between discontinuous and continuous modes is reached when the inductor current falls to zero exactly at the end of the commutation cycle. Using the notations of figure, this corresponds to:

$$DT + \delta T = T$$
$$\Rightarrow D + \delta = 1$$

Therefore, the output current (equal to the average inductor current) at the limit between discontinuous and continuous modes is:

$$I_{o_{lim}} = \frac{I_{L_{max}}}{2}(D + \delta) = \frac{I_{L_{max}}}{2}$$

Substituting I_{Lmax} by its value:

$$I_{o_{lim}} = \frac{V_i - V_o}{2L}DT$$

On the limit between the two modes, the output voltage obeys both the expressions given respectively in the continuous and the discontinuous sections. In particular, the former is:

$$V_o = DV_i$$

So $I_{o_{lim}}$ can be written as:

$$I_{o_{lim}} = \frac{V_i(1-D)}{2L}DT$$

Let's now introduce two more notations:

- The normalized voltage, defined by $|V_o| = \frac{V_o}{V_i}$. It is zero when $V_o = 0$, and 1 when $V_o = V_i$

- The normalized current, defined by $|I_o| = \frac{L}{TV_i}I_o$. The term $\frac{TV_i}{L}$ is equal to the maximum increase of the inductor current during a cycle; i.e., the increase of the inductor current with a duty cycle D=1. So, in steady state operation of the converter, this means that $|I_o|$ equals 0 for no output current, and 1 for the maximum current the converter can deliver.

Using these notations, we have:

- In continuous mode:

$$|V_o| = D$$

- In discontinuous mode:

$$|V_o| = \frac{1}{\frac{2LI_o}{D^2V_iT}+1}$$

$$= \frac{1}{\frac{2|I_o|}{D^2}+1}$$

$$= \frac{D^2}{2|I_o|+D^2}$$

The current at the limit between continuous and discontinuous mode is:

$$I_{o_{lim}} = \frac{V_i}{2L}D(1-D)T$$

$$= \frac{I_o}{2|I_o|}D(1-D)$$

Therefore, the locus of the limit between continuous and discontinuous modes is given by:

$$\frac{(1-D)D}{2|I_o|} = 1$$

These expressions have been plotted in figure. From this, it can be deduced that in continuous mode, the output voltage does only depend on the duty cycle, whereas it is far more complex in the discontinuous mode. This is important from a control point of view.

On the circuit level, the detection of the boundary between CCM and DCM are usually provided by an inductor current sensing, requiring high accuracy and fast detectors as:

Non-ideal Circuit

The previous study was conducted with the following assumptions:

- The output capacitor has enough capacitance to supply power to the load (a simple resistance) without any noticeable variation in its voltage.

- The voltage drop across the diode when forward biased is zero.

- No commutation losses in the switch nor in the diode.

These assumptions can be fairly far from reality, and the imperfections of the real components can have a detrimental effect on the operation of the converter.

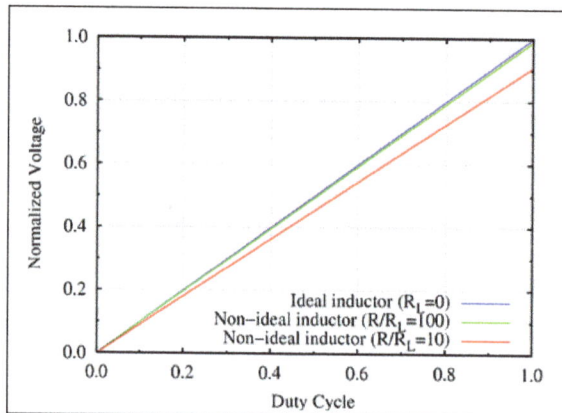

Evolution of the output voltage of a buck converter with the duty cycle when the parasitic resistance of the inductor increases.

Output Voltage Ripple (Continuous Mode)

Output voltage ripple is the name given to the phenomenon where the output voltage rises during the On-state and falls during the Off-state. Several factors contribute to this including, but not limited to, switching frequency, output capacitance, inductor, load and any current limiting features of the control circuitry. At the most basic level the output voltage will rise and fall as a result of the output capacitor charging and discharging:

$$dV_o = \frac{idT}{C}$$

We can best approximate output ripple voltage by shifting the output current versus time waveform (continuous mode) down so that the average output current is along the time axis. When we do this, we see the AC current waveform flowing into and out of the output capacitor (sawtooth waveform). We note that Vc-min (where Vc is the capacitor voltage) occurs at t-on/2 (just after capacitor has discharged) and Vc-max at t-off/2. By integrating Idt (= dQ ; as I = dQ/dt, C = Q/V so dV = dQ/C) under the output current waveform through writing output ripple voltage as dV = Idt/C we integrate the area above the axis to get the peak-to-peak ripple voltage as: delta V = delta I * T /8C (where delta I is the peak-to-peak ripple current and T is the time period of ripple; see Talk tab for details if you can't graphically work out the areas here. A full explanation is given there.) We note from basic AC circuit theory that our ripple voltage should be roughly sinusoidal: capacitor impedance times ripple current peak-to-peak value, or delta V = delta I / (2 * omega * C) where omega = 2 * pi * f, f is the ripple frequency, and f = 1/T, T the ripple period. This gives: delta V = delta I * T / (2 * pi * C), and we compare to this value to confirm the above in that we have a factor of 8 vs a factor of ~ 6.3 from basic AC circuit theory for a sinusoid.

During the Off-state, the current in this equation is the load current. In the On-state the current is the difference between the switch current (or source current) and the load current. The duration of time (dT) is defined by the duty cycle and by the switching frequency.

For the On-state:

$$dT_{on} = DT = \frac{D}{f}$$

For the Off-state:

$$dT_{off} = (1-D)T = \frac{1-D}{f}$$

Qualitatively, as the output capacitor or switching frequency increase, the magnitude of the ripple decreases. Output voltage ripple is typically a design specification for the power supply and is selected based on several factors. Capacitor selection is normally determined based on cost, physical size and non-idealities of various capacitor types. Switching frequency selection is typically determined based on efficiency requirements, which tends to decrease at higher operating frequencies. Higher switching frequency can also raise EMI concerns.

Output voltage ripple is one of the disadvantages of a switching power supply, and can also be a measure of its quality.

Effects of non-ideality on the Efficiency

A simplified analysis of the buck converter, does not account for non-idealities of the circuit components nor does it account for the required control circuitry. Power losses due to the control circuitry are usually insignificant when compared with the losses in the power devices (switches, diodes, inductors, etc.) The non-idealities of the power devices account for the bulk of the power losses in the converter.

Both static and dynamic power losses occur in any switching regulator. Static power losses include I^2R (conduction) losses in the wires or PCB traces, as well as in the switches and inductor, as in any electrical circuit. Dynamic power losses occur as a result of switching, such as the charging and discharging of the switch gate, and are proportional to the switching frequency.

It is useful to begin by calculating the duty cycle for a non-ideal buck converter, which is:

$$D = \frac{V_o + (V_{sw,sync} + V_L)}{V_i - V_{sw} + V_{sw,sync}}$$

where,

V_{sw} is the voltage drop on the power switch,

$V_{sw,sync}$ is the voltage drop on the synchronous switch or diode, and

V_L is the voltage drop on the inductor.

The voltage drops are all static power losses which are dependent primarily on DC current, and can therefore be easily calculated. For a diode drop, V_{sw} and $V_{sw,sync}$ may already be known, based on the properties of the selected device.

$$V_{sw} = I_{sw}R_{on} = DI_oR_{on}$$
$$V_{sw,sync} = I_{sw,sync}R_{on} = (1-D)I_oR_{on}$$
$$V_L = I_LR_{DC}$$

where,

R_{on} is the ON-resistance of each switch, and

R_{DC} is the DC resistance of the inductor.

The duty cycle equation is somewhat recursive. A rough analysis can be made by first calculating the values V_{sw} and $V_{sw,sync}$ using the ideal duty cycle equation.

For a MOSFET voltage drop, a common approximation is to use R_{DSon} from the MOSFET's datasheet in Ohm's Law, $V = I_{DS}R_{DSon(sat)}$. This approximation is acceptable because the MOSFET is in the linear state, with a relatively constant drain-source resistance.

This approximation is only valid at relatively low V_{DS} values. For more accurate calculations, MOSFET datasheets contain graphs on the V_{DS} and I_{DS} relationship at multiple V_{GS} values. Observe V_{DS} at the V_{GS} and I_{DS} which most closely match what is expected in the buck converter.

In addition, power loss occurs as a result of leakage currents. This power loss is simply,

$$P_{leakage} = I_{leakage}V$$

where:

- $I_{leakage}$ is the leakage current of the switch, and
- V is the voltage across the switch.

Dynamic power losses are due to the switching behavior of the selected pass devices (MOSFETs, power transistors, IGBTs, etc.). These losses include turn-on and turn-off switching losses and switch transition losses.

Switch turn-on and turn-off losses are easily lumped together as:

$$P_{SW} = \frac{VI_o(t_{rise} + t_{fall})}{6T}$$

where,

V is the voltage across the switch while the switch is off,

t_{rise} and t_{fall} are the switch rise and fall times,

T is the switching period.

but this does not take into account the parasitic capacitance of the MOSFET which makes the *Miller plate*. Then, the switch losses will be more like:

$$P_{SW} = \frac{VI_o(t_{rise} + t_{fall})}{2T}$$

When a MOSFET is used for the lower switch, additional losses may occur during the time between the turn-off of the high-side switch and the turn-on of the low-side switch, when the body diode of the low-side MOSFET conducts the output current. This time, known as the non-overlap time, prevents "shootthrough", a condition in which both switches are simultaneously turned on. The onset of shootthrough generates severe power loss and heat. Proper selection of non-overlap time must balance the risk of shootthrough with the increased power loss caused by conduction of the body diode. Many MOSFET based buck converters also include a diode to aid the lower MOSFET body diode with conduction during the non-overlap time. When a diode is used exclusively for the lower switch, diode forward turn-on time can reduce efficiency and lead to voltage overshoot.

Power loss on the body diode is also proportional to switching frequency and is:

$$P_{D,body} = V_F I_o t_{no} f_{SW}$$

where,

V_F is the forward voltage of the body diode, and

t_{no} is the selected non-overlap time.

Finally, power losses occur as a result of the power required to turn the switches on and off. For MOSFET switches, these losses are dominated by the energy required to charge and discharge the capacitance of the MOSFET gate between the threshold voltage and the selected gate voltage. These switch transition losses occur primarily in the gate driver, and can be minimized by selecting MOSFETs with low gate charge, by driving the MOSFET gate to a lower voltage (at the cost of increased MOSFET conduction losses), or by operating at a lower frequency.

$$P_{Gdrive} = Q_G V_{GS} f_{SW}$$

where,

Q_G is the gate charge of the selected MOSFET, and

V_{GS} is the peak gate-source voltage.

For N-MOSFETs, the high-side switch must be driven to a higher voltage than V_i. To achieve this, MOSFET gate drivers typically feed the MOSFET output voltage back into the gate driver. The gate driver then adds its own supply voltage to the MOSFET output voltage when driving the high-side MOSFETs to achieve a V_{GS} equal to the gate driver supply voltage. Because the low-side V_{GS} is the gate driver supply voltage, this results in very similar V_{GS} values for high-side and low-side MOSFETs.

A complete design for a buck converter includes a tradeoff analysis of the various power losses. Designers balance these losses according to the expected uses of the finished design. A converter expected to have a low switching frequency does not require switches with low gate transition losses; a converter operating at a high duty cycle requires a low-side switch with low conduction losses.

Specific Structures

Synchronous Rectification

A synchronous buck converter is a modified version of the basic buck converter circuit topology in which the diode, D, is replaced by a second switch, S_2. This modification is a tradeoff between increased cost and improved efficiency.

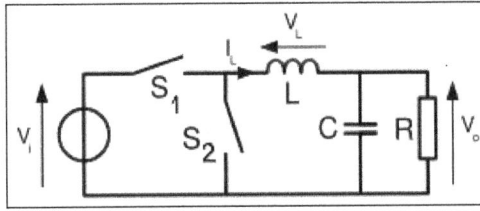

Simplified schematic of a synchronous converter, in which D is replaced by a second switch, S_2.

In a standard buck converter, the flyback diode turns on, on its own, shortly after the switch turns off, as a result of the rising voltage across the diode. This voltage drop across the diode results in a power loss which is equal to:

$$P_D = V_D(1-D)I_o$$

where,

V_D is the voltage drop across the diode at the load current I_o,

D is the duty cycle, and

I_o is the load current.

By replacing the diode with a switch selected for low loss, the converter efficiency can be improved. For example, a MOSFET with very low R_{DSon} might be selected for S_2, providing power loss on switch S_2 which is:

$$P_{S_2} = I_o^2 R_{DSon}(1-D)$$

In both cases, power loss is strongly dependent on the duty cycle, D. Power loss on the freewheeling diode or lower switch will be proportional to its on-time. Therefore, systems designed for low duty cycle operation will suffer from higher losses in the freewheeling diode or lower switch, and for such systems it is advantageous to consider a synchronous buck converter design.

Consider a computer power supply, where the input is 5 V, the output is 3.3 V, and the load current is 10 A. In this case, the duty cycle will be 66% and the diode would be on for 34% of the time. A typical diode with forward voltage of 0.7 V would suffer a power loss of 2.38 W. A well-selected MOSFET with R_{DSon} of 0.015 Ω, however, would waste only 0.51 W in conduction loss. This translates to improved efficiency and reduced heat generation.

Another advantage of the synchronous converter is that it is bi-directional, which lends itself to applications requiring regenerative braking. When power is transferred in the "reverse" direction, it acts much like a boost converter.

The advantages of the synchronous buck converter do not come without cost. First, the lower switch typically costs more than the freewheeling diode. Second, the complexity of the converter is vastly increased due to the need for a complementary-output switch driver.

Such a driver must prevent both switches from being turned on at the same time, a fault known as "shootthrough". The simplest technique for avoiding shootthrough is a time delay between the turn-off of S_1 to the turn-on of S_2, and vice versa. However, setting this time delay long enough to ensure that S_1 and S_2 are never both on will itself result in excess power loss. An improved technique for preventing this condition is known as adaptive "non-overlap" protection, in which the voltage at the switch node (the point where S_1, S_2 and L are joined) is sensed to determine its state. When the switch node voltage passes a preset threshold, the time delay is started. The driver can thus adjust to many types of switches without the excessive power loss this flexibility would cause with a fixed non-overlap time.

Multiphase Buck

The multiphase buck converter is a circuit topology where basic buck converter circuits are placed in parallel between the input and load. Each of the n "phases" is turned on at equally spaced intervals over the switching period. This circuit is typically used with the synchronous buck topology.

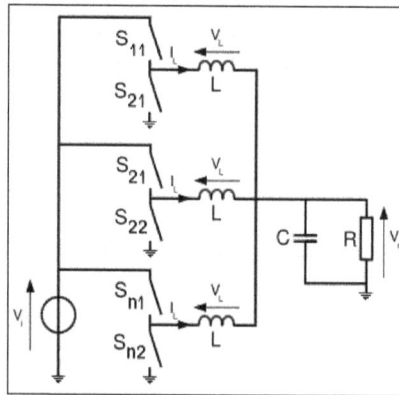

Schematic of a generic synchronous n-phase buck converter.

This type of converter can respond to load changes as quickly as if it switched n times faster, without the increase in switching losses that would cause. Thus, it can respond to rapidly changing loads, such as modern microprocessors.

There is also a significant decrease in switching ripple. Not only is there the decrease due to the increased effective frequency, but any time that n times the duty cycle is an integer, the switching ripple goes to 0; the rate at which the inductor current is increasing in the phases which are switched on exactly matches the rate at which it is decreasing in the phases which are switched off.

Closeup picture of a multiphase CPU power supply for an AMD Socket 939 processor.
The three phases of this supply can be recognized by the three black toroidal inductors
in the foreground. The smaller inductor below the heat sink is part of an input filter.

Another advantage is that the load current is split among the n phases of the multiphase converter. This load splitting allows the heat losses on each of the switches to be spread across a larger area.

This circuit topology is used in computer motherboards to convert the 12 V_{DC} power supply to a lower voltage (around 1 V), suitable for the CPU. Modern CPU power requirements can exceed 200 W, can change very rapidly, and have very tight ripple requirements, less than 10 mV. Typical motherboard power supplies use 3 or 4 phases.

One major challenge inherent in the multiphase converter is ensuring the load current is balanced evenly across the n phases. This current balancing can be performed in a number of ways. Current can be measured "losslessly" by sensing the voltage across the inductor or the lower switch (when it is turned on). This technique is considered lossless because it relies on resistive losses inherent in the buck converter topology. Another technique is to insert a small resistor in the circuit and measure the voltage across it. This approach is more accurate and adjustable, but incurs several costs—space, efficiency and money.

Finally, the current can be measured at the input. Voltage can be measured losslessly, across the upper switch, or using a power resistor, to approximate the current being drawn. This approach is technically more challenging, since switching noise cannot be easily filtered out. However, it is less expensive than emplacing a sense resistor for each phase.

Efficiency Factors

Conduction losses that depend on load:

- Resistance when the transistor or MOSFET switch is conducting.

- Diode forward voltage drop (usually 0.7 V or 0.4 V for schottky diode).

- Inductor winding resistance.

- Capacitor equivalent series resistance.

Switching losses:

- Voltage-Ampere overlap loss.
- $\text{Frequency}_{\text{switch}} {}^*CV^2$ loss.
- Reverse latence loss.
- Losses due driving MOSFET gate and controller consumption.
- Transistor leakage current losses, and controller standby consumption.

Impedance Matching

A buck converter can be used to maximize the power transfer through the use of impedance matching. An application of this is in a maximum power point tracker commonly used in photovoltaic systems.

By the equation for electric power:

$$V_o I_o = \eta V_i I_i$$

where,

V_o is the output voltage,

I_o is the output current,

η is the power efficiency (ranging from 0 to 1),

V_i is the input voltage,

I_i is the input current.

By Ohm's law:

$$I_o = \frac{V_o}{Z_o}$$

$$I_i = \frac{V_i}{Z_i}$$

where,

Z_o is the output impedance,

Z_i is the input impedance.

Substituting these expressions for I_o and I_i into the power equation yields:

$$\frac{V_o^2}{Z_o} = \frac{\eta V_i^2}{Z_i}$$

As was previously shown for the continuous mode, (where $I_L > 0$):

$$V_o = DV_i$$

where,

> D is the duty cycle.

Substituting this equation for V_o into the previous equation, yields:

$$\frac{\left(DV_i\right)^2}{Z_o} = \frac{\eta V_i^2}{Z_i}$$

which reduces to:

$$\frac{D^2}{Z_o} = \frac{\eta}{Z_i}$$

and finally:

$$D = \sqrt{\frac{\eta Z_o}{Z_i}}$$

This shows that it is possible to adjust the impedance ratio by adjusting the duty cycle. This is particularly useful in applications where the impedance(s) are dynamically changing.

Application

The buck is widely used in low power consumption small electronics to step-down from 24/12V down to 5V. They are sold as a small finish product chip for well less than $1 USD having about 95% efficiency.

Boost Converter

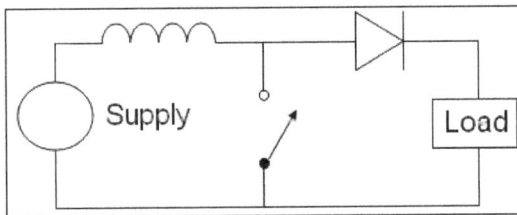

The basic schematic of a boost converter. The switch is typically a MOSFET, IGBT, or BJT.

A boost converter (step-up converter) is a DC-to-DC power converter that steps up voltage (while stepping down current) from its input (supply) to its output (load). It is a class of switched-mode power supply (SMPS) containing at least two semiconductors (a diode and a transistor) and at least one energy storage element: a capacitor, inductor, or the two in combination. To reduce voltage ripple, filters made of capacitors (sometimes in combination with inductors) are normally added to such a converter's output (load-side filter) and input (supply-side filter).

Power for the boost converter can come from any suitable DC source, such as batteries, solar panels, rectifiers, and DC generators. A process that changes one DC voltage to a different DC voltage is called DC to DC conversion. A boost converter is a DC to DC converter with an output voltage greater than the source voltage. A boost converter is sometimes called a step-up converter since it "steps up" the source voltage. Since power ($P = VI$) must be conserved, the output current is lower than the source current.

Applications

Boost converter from a TI calculator, generating 9 V from 2.4 V provided by two AA rechargeable cells.

Battery power systems often stack cells in series to achieve higher voltage. However, sufficient stacking of cells is not possible in many high voltage applications due to lack of space. Boost converters can increase the voltage and reduce the number of cells. Two battery-powered applications that use boost converters are used in hybrid electric vehicles (HEV) and lighting systems.

The NHW20 model Toyota Prius HEV uses a 500 V motor. Without a boost converter, the Prius would need nearly 417 cells to power the motor. However, a Prius actually uses only 168 cells and boosts the battery voltage from 202 V to 500 V. Boost converters also power devices at smaller scale applications, such as portable lighting systems. A white LED typically requires 3.3 V to emit light, and a boost converter can step up the voltage from a single 1.5 V alkaline cell to power the lamp.

An unregulated boost converter is used as the voltage increase mechanism in the circuit known as the 'Joule thief'. This circuit topology is used with low power battery applications, and is aimed at the ability of a boost converter to 'steal' the remaining energy in a battery. This energy would otherwise be wasted since the low voltage of a nearly depleted battery makes it unusable for a normal load. This energy would otherwise remain untapped because many applications do not allow enough current to flow through a load when voltage decreases. This voltage decrease occurs as batteries become depleted, and is a characteristic of the ubiquitous alkaline battery. Since the equation for power is ($P = V^2 / R$), and R tends to be stable, power available to the load goes down significantly as voltage decreases.

Circuit analysis

Operation

The key principle that drives the boost converter is the tendency of an inductor to resist changes in current by creating and destroying a magnetic field. In a boost converter, the output voltage is always higher than the input voltage. A schematic of a boost power stage is shown in figure.

- When the switch is closed, current flows through the inductor in clockwise direction and the inductor stores some energy by generating a magnetic field. Polarity of the left side of the inductor is positive.

- When the switch is opened, current will be reduced as the impedance is higher. The magnetic field previously created will be destroyed to maintain the current towards the load. Thus the polarity will be reversed (meaning the left side of the inductor will become negative). As a result, two sources will be in series causing a higher voltage to charge the capacitor through the diode D.

If the switch is cycled fast enough, the inductor will not discharge fully in between charging stages, and the load will always see a voltage greater than that of the input source alone when the switch is opened. Also while the switch is opened, the capacitor in parallel with the load is charged to this combined voltage. When the switch is then closed and the right hand side is shorted out from the left hand side, the capacitor is therefore able to provide the voltage and energy to the load. During this time, the blocking diode prevents the capacitor from discharging through the switch. The switch must of course be opened again fast enough to prevent the capacitor from discharging too much.

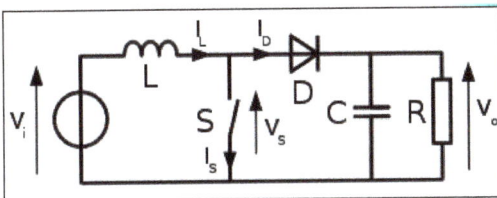

Boost converter schematic.

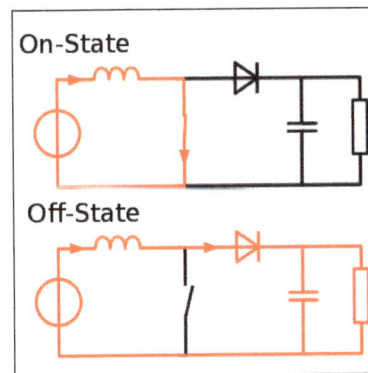

The two current paths of a boost converter, depending on the state of the switch S.

The basic principle of a Boost converter consists of 2 distinct states:

- In the On-state, the switch S is closed, resulting in an increase in the inductor current.

- In the Off-state, the switch is open and the only path offered to inductor current is through the flyback diode D, the capacitor C and the load R. This results in transferring the energy accumulated during the On-state into the capacitor.

The input current is the same as the inductor current as can be seen in figure. So it is not discontinuous as in the buck converter and the requirements on the input filter are relaxed compared to a buck converter.

Continuous Mode

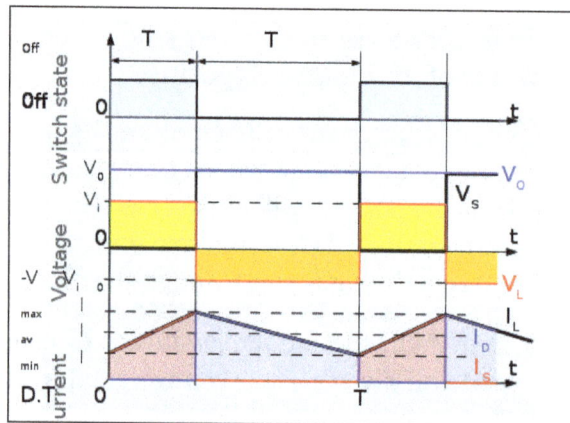

Waveforms of current and voltage in a boost converter operating in continuous mode.

When a boost converter operates in continuous mode, the current through the inductor (I_L) never falls to zero. Figure shows the typical waveforms of currents and voltages in a converter operating in this mode.

In the steady state, the DC (average) voltage across the inductor must be zero so that after each cycle the inductor returns the same state, because voltage across the inductor is proportional to rate of change of current through it. Note in figure that the left hand side of L is at V_i and the right hand side of L sees the V_s voltage waveform from figure. The average value of V_s is $(1-D)V_o$ where D is the duty cycle of the waveform driving the switch. From this we get the ideal transfer function:

$$V_i = (1-D)V_o$$

Or

$$V_o / V_i = 1/(1-D).$$

We get the same result from a more detailed analysis as follows: The output voltage can be calculated as follows, in the case of an ideal converter (i.e. using components with an ideal behaviour) operating in steady conditions:

During the On-state, the switch S is closed, which makes the input voltage (V_i) appear across the inductor, which causes a change in current (I_L) flowing through the inductor during a time period (t) by the formula:

$$\frac{\Delta I_L}{\Delta t} = \frac{V_i}{L}$$

Where, L is the inductor value.

At the end of the On-state, the increase of I_L is therefore:

$$\Delta I_{L_{On}} = \frac{1}{L}\int_0^{DT} V_i dt = \frac{DT}{L}V_i$$

D is the duty cycle. It represents the fraction of the commutation period T during which the switch is On. Therefore, D ranges between 0 (S is never on) and 1 (S is always on).

During the Off-state, the switch S is open, so the inductor current flows through the load. If we consider zero voltage drop in the diode, and a capacitor large enough for its voltage to remain constant, the evolution of I_L is:

$$V_o - V_i = L\frac{dI_L}{dt}$$

Therefore, the variation of I_L during the Off-period is:

$$\Delta I_{L_{Off}} = \int_{DT}^{T}\frac{(V_i - V_o)dt}{L} = \frac{(V_i - V_o)(1-D)T}{L}$$

As we consider that the converter operates in steady-state conditions, the amount of energy stored in each of its components has to be the same at the beginning and at the end of a commutation cycle. In particular, the energy stored in the inductor is given by:

$$E = \frac{1}{2}LI_L^2$$

So, the inductor current has to be the same at the start and end of the commutation cycle. This means the overall change in the current (the sum of the changes) is zero:

$$\Delta I_{L_{On}} + \Delta I_{L_{Off}} = 0$$

Substituting $\Delta I_{L_{On}} \Delta I_{L_{On}}$ and $\Delta I_{L_{Off}} \Delta I_{L_{Off}}$

$$\Delta I_{L_{On}} + \Delta I_{L_{Off}} = \frac{V_i DT}{L} + \frac{(V_i - V_o)(1-D)T}{L} = 0$$

This can be written as:

$$\frac{V_o}{V_i} = \frac{1}{1-D}$$

The above equation shows that the output voltage is always higher than the input voltage (as the duty cycle goes from 0 to 1), and that it increases with D, theoretically to infinity as D approaches 1. This is why this converter is sometimes referred to as a step-up converter.

Rearranging the equation reveals the duty cycle to be:

$$D = 1 - \frac{V_i}{V_o}$$

Discontinuous Mode

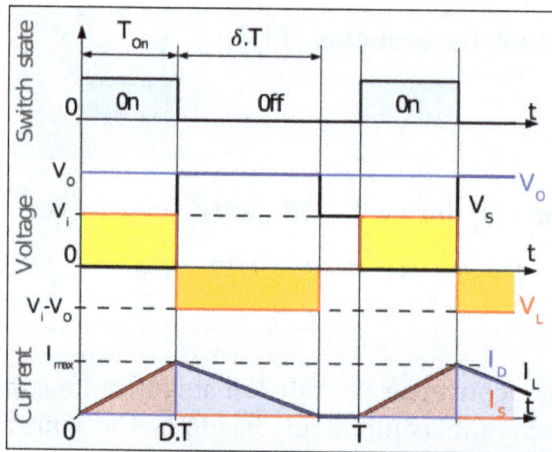

Waveforms of current and voltage in a boost converter operating in discontinuous mode.

If the ripple amplitude of the current is too high, the inductor may be completely discharged before the end of a whole commutation cycle. This commonly occurs under light loads. In this case, the current through the inductor falls to zero during part of the period.

Although the difference is slight, it has a strong effect on the output voltage equation. The voltage gain can be calculated as follows:

As the inductor current at the beginning of the cycle is zero, its maximum value,

$I_{L_{Max}}$ (at $t = DT$) is

$$I_{L_{Max}} = \frac{V_i DT}{L}$$

During the off-period, I_L falls to zero after δT:

$$I_{L_{Max}} + \frac{(V_i - V_o)\delta T}{L} = 0$$

Using the two previous equations, δ is:

$$\delta = \frac{V_i D}{V_o - V_i}$$

The load current I_o is equal to the average diode current (I_D). As can be seen on figure, the diode current is equal to the inductor current during the off-state. Therefore, the output current can be written as:

$$I_o = \overline{I}_D = \frac{I_{L_{max}}}{2}\delta$$

Replacing I_{Lmax} and δ by their respective expressions yields:

$$I_o = \frac{V_i DT}{2L} \cdot \frac{V_i D}{V_o - V_i} = \frac{V_i^2 D^2 T}{2L(V_o - V_i)}$$

Therefore, the output voltage gain can be written as follows:

$$\frac{V_o}{V_i} = 1 + \frac{V_i D^2 T}{2LI_o}$$

Compared to the expression of the output voltage gain for continuous mode, this expression is much more complicated. Furthermore, in discontinuous operation, the output voltage gain not only depends on the duty cycle (D), but also on the inductor value (L), the input voltage (V_i), the commutation period (T) and the output current (I_o).

Substituting $I_o = V_o/R$ into the equation (R is the load), the output voltage gain can be rewritten as:

$$\frac{V_o}{V_i} = \frac{1 + \sqrt{1 + \frac{4D^2}{K}}}{2}$$

where, $K = \frac{2L}{RT}$

Ćuk Converter

The Ćuk converter is a type of DC/DC converter that has an output voltage magnitude that is either greater than or less than the input voltage magnitude. It is essentially a boost converter followed by a buck converter with a capacitor to couple the energy.

Similar to the buck–boost converter with inverting topology, the output voltage of non-isolated Ćuk is typically also inverting, and can be lower or higher than the input. It uses a capacitor as its main energy-storage component, unlike most other types of converters which use an inductor. It is named after Slobodan Ćuk of the California Institute of Technology, who first presented the design.

Non-isolated Ćuk Converter

There are variations on the basic Ćuk converter. For example, the coils may share single magnetic core, which drops the output ripple, and adds efficiency. Because the power transfer flows continuously via the capacitor, this type of switcher has minimized EMI radiation. The Ćuk converter allows energy to flow bidirectionally by using a diode and a switch.

Operating Principle

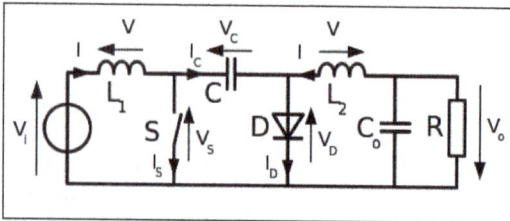

Schematic of a non-isolated Ćuk converter.

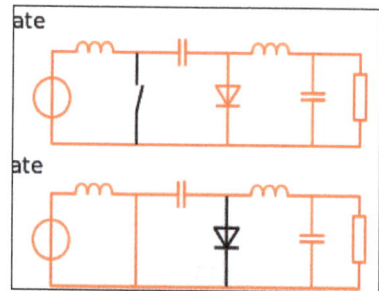

The two operating states of a non-isolated Ćuk converter.

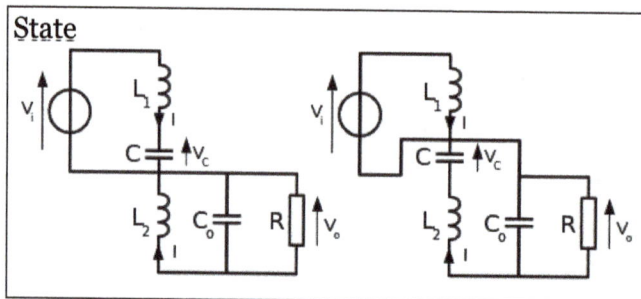

The two operating states of a non-isolated Ćuk converter. In this figure, the diode and the switch are either replaced by a short circuit when they are on or by an open circuit when they are off. It can be seen that when in the off-state, the capacitor C is being charged by the input source through the inductor L1. When in the on-state, the capacitor C transfers the energy to the output capacitor through the inductance L2.

A non-isolated Ćuk converter comprises two inductors, two capacitors, a switch (usually a transistor), and a diode. Its schematic can be seen in figure. It is an inverting converter, so the output voltage is negative with respect to the input voltage.

The capacitor C is used to transfer energy and is connected alternately to the input and to the output of the converter via the commutation of the transistor and the diode.

The two inductors L_1 and L_2 are used to convert respectively the input voltage source (Vi) and the output voltage source (Co) into current sources. At a short time scale an inductor can be considered as a current source as it maintains a constant current. This conversion is necessary because if the capacitor were connected directly to the voltage source, the current would be limited only by the parasitic resistance, resulting in high energy loss. Charging a capacitor with a current source (the inductor) prevents resistive current limiting and its associated energy loss.

As with other converters (buck converter, boost converter, buck–boost converter) the Ćuk converter can either operate in continuous or discontinuous current mode. However, unlike these converters, it can also operate in discontinuous voltage mode (the voltage across the capacitor drops to zero during the commutation cycle).

Continuous Mode

In steady state, the energy stored in the inductors has to remain the same at the beginning and at the end of a commutation cycle. The energy in an inductor is given by:

$$E = \frac{1}{2}LI^2$$

This implies that the current through the inductors has to be the same at the beginning and the end of the commutation cycle. As the evolution of the current through an inductor is related to the voltage across it:

$$V_L = L\frac{dI}{dt}$$

it can be seen that the average value of the inductor voltages over a commutation period have to be zero to satisfy the steady-state requirements.

If we consider that the capacitors C and C_o are large enough for the voltage ripple across them to be negligible, the inductor voltages become:

- In the off-state, inductor L_1 is connected in series with V_i and C. Therefore

$V_{L1} = V_i - V_C$ As the diode D is forward biased (we consider zero voltage drop), L_2 is directly connected to the output capacitor. Therefore $V_{L2} = V_o$.

- In the on-state, inductor L_1 is directly connected to the input source. Therefore $V_{L1} = V_i$ Inductor L_2 is connected in series with C and the output capacitor, so $V_{L2} = V_o + V_C$.

The converter operates in on state from t=0 to t=D·T (D is the duty cycle), and in off state from D·T to T (that is, during a period equal to (1-D)·T). The average values of V_{L1} and V_{L2} are therefore:

$$\bar{V}_{L1} = D \cdot V_i + (1-D) \cdot (V_i - V_C) = (V_i - (1-D) \cdot V_C)$$

$$\bar{V}_{L2} = D(V_o + V_C) + (1-D) \cdot V_o = (V_o + D \cdot V_C)$$

As both average voltage have to be zero to satisfy the steady-state conditions, using the last equation we can write:

$$V_C = -\frac{V_o}{D}$$

So the average voltage across L_1 becomes:

$$\bar{V}_{L1} = \left(V_i + (1-D) \cdot \frac{V_o}{D} \right) = 0$$

Which can be written as:

$$\frac{V_o}{V_i} = \frac{-D}{1-D}$$

It can be seen that this relation is the same as that obtained for the buck–boost converter.

Discontinuous Mode

Like all DC/DC converters Ćuk converters rely on the ability of the inductors in the circuit to provide continuous current, in much the same way a capacitor in a rectifier filter provides continuous voltage. If this inductor is too small or below the "critical inductance", then the inductor current slope will be discontinuous where the current goes to zero. This state of operation is usually not studied in much depth as it is generally not used beyond a demonstrating of why the minimum inductance is crucial, although it may occur when maintaining a standby voltage at a much lower current than the converter was designed for.

The minimum inductance is given by:

$$L_1 min = \frac{(1-D)^2 R}{2Df_s}$$

Where, f_s is the switching frequency.

Isolated Ćuk Converter

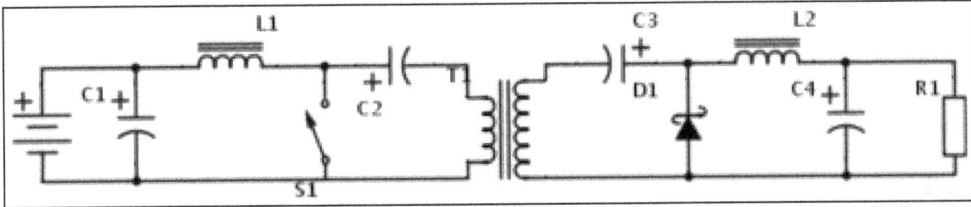

Isolated Ćuk converter with gapless AC transformer in the middle.

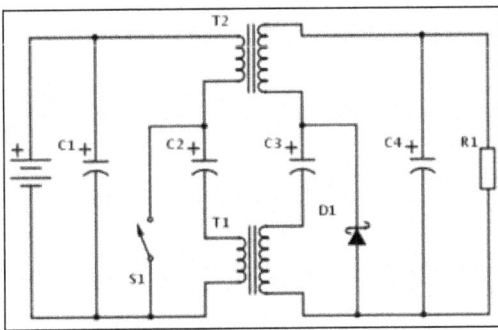

Coupled inductor isolated Ćuk converter.

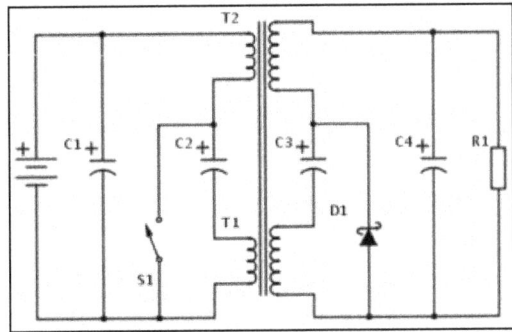

Integrated magnetics Ćuk converter.

The Ćuk converter can be made in an isolated kind. An AC transformer and an additional capacitor must be added.

Because the isolated Ćuk converter is isolated, the output-voltage polarity can be chosen freely.

As the non-isolated Ćuk converter, the isolated Ćuk converter can have an output voltage magnitude that is either greater than or less than the input voltage magnitude, even with a 1:1 AC transformer.

Related Structures

Inductor Coupling

Instead of using two discrete inductor components, many designers implement a coupled inductor Ćuk converter, using a single magnetic component that includes both inductors on the same core. The transformer action between the inductors inside that component gives a coupled inductor Ćuk converter with lower output ripple than a Ćuk converter using two independent discrete inductor components.

Zeta Converter

A zeta converter provides an ouput voltage that is the opposite of the output voltage of a Ćuk converter.

Single-ended Primary-inductance Converter (SEPIC)

A SEPIC converter is able to step-up or step-down the voltage.

Buck–boost Converter

The buck–boost converter is a type of DC-to-DC converter that has an output voltage magnitude that is either greater than or less than the input voltage magnitude. It is equivalent to a flyback converter using a single inductor instead of a transformer.

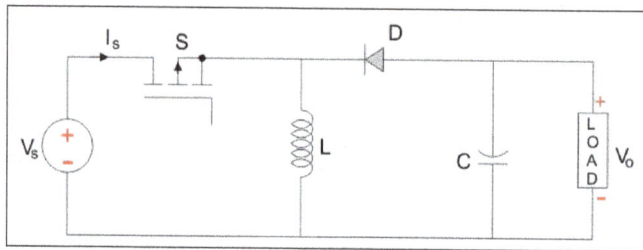

The basic schematic of an inverting buck–boost converter.

Two different topologies are called buck–boost converter. Both of them can produce a range of output voltages, ranging from much larger (in absolute magnitude) than the input voltage, down to almost zero.

The Inverting Topology

The output voltage is of the opposite polarity than the input. This is a switched-mode power supply with a similar circuit topology to the boost converter and the buck converter. The output voltage is adjustable based on the duty cycle of the switching transistor. One possible drawback of this converter is that the switch does not have a terminal at ground; this complicates the driving circuitry. However, this drawback is of no consequence if the power supply is isolated from the load circuit (if, for example, the supply is a battery) because the supply and diode polarity can simply be reversed. When they can be reversed, the switch can be on either the ground side or the supply side.

A Buck (Step-down) Converter Combined with a Boost (Step-up) Converter

The output voltage is typically of the same polarity of the input, and can be lower or higher than the input. Such a non-inverting buck-boost converter may use a single inductor which is used for both the buck inductor mode and the boost inductor mode, using switches instead of diodes, sometimes called a "four-switch buck-boost converter", it may use multiple inductors but only a single switch as in the SEPIC and Ćuk topologies.

Principle of Operation of the Inverting Topology

The basic principle of the inverting buck–boost converter is fairly simple:

- While in the On-state, the input voltage source is directly connected to the inductor (L). This results in accumulating energy in L. In this stage, the capacitor supplies energy to the output load.

- While in the Off-state, the inductor is connected to the output load and capacitor, so energy is transferred from L to C and R.

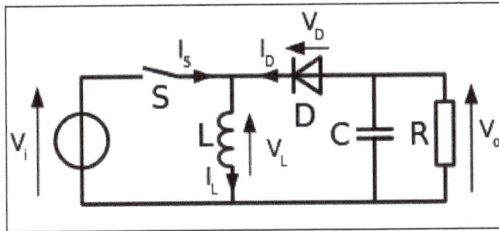

Schematic of a buck–boost converter.

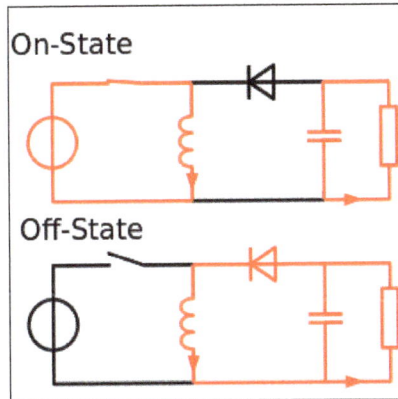

The two operating states of a buck–boost converter: When the switch is turned on, the input voltage source supplies current to the inductor, and the capacitor supplies current to the resistor (output load). When the switch is opened, the inductor supplies current to the load via the diode D.

Compared to the buck and boost converters, the characteristics of the inverting buck–boost converter are mainly:

- Polarity of the output voltage is opposite to that of the input.

- The output voltage can vary continuously from 0 to $-\infty$ (for an ideal converter). The output voltage ranges for a buck and a boost converter are respectively v_i to 0 and v_i ∞.

Conceptual Overview

Like the buck and boost converters, the operation of the buck-boost is best understood in terms of the inductor's "reluctance" to allow rapid change in current. From the initial state in which nothing is charged and the switch is open, the current through the inductor is zero. When the switch is first closed, the blocking diode prevents current from flowing into the right hand side of the circuit, so it must all flow through the inductor. However, since the inductor doesn't allow rapid current change, it will initially keep the current low by dropping most of the voltage provided by the source. Over time, the inductor will allow the current to slowly increase by decreasing its voltage drop. Also during this time, the inductor will store energy in the form of a magnetic field.

Continuous Mode

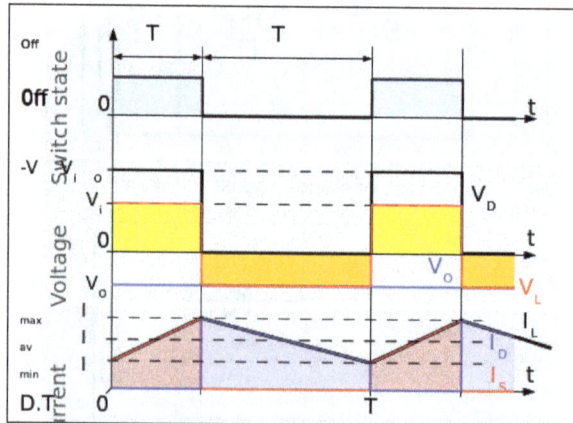

Waveforms of current and voltage in a buck–boost converter operating in continuous mode.

If the current through the inductor L never falls to zero during a commutation cycle, the converter is said to operate in continuous mode. The current and voltage waveforms in an ideal converter can be seen in figure.

From t = 0 to t = DT, the converter is in On-State, so the switch S is closed. The rate of change in the inductor current (I_L) is therefore given by,

$$\frac{dI_L}{dt} = \frac{V_i}{L}$$

At the end of the On-state, the increase of I_L is therefore:

$$\Delta I_{L_{On}} = \int_0^{DT} dI_L = \int_0^{DT} \frac{V_i}{L} dt = \frac{V_i DT}{L}$$

D is the duty cycle. It represents the fraction of the commutation period T during which the switch is On. Therefore D ranges between 0 (S is never on) and 1 (S is always on).

During the Off-state, the switch S is open, so the inductor current flows through the load. If we assume zero voltage drop in the diode, and a capacitor large enough for its voltage to remain constant, the evolution of I_L is:

$$\frac{dI_L}{dt} = \frac{V_o}{L}$$

Therefore, the variation of I_L during the Off-period is:

$$\Delta I_{L_{Off}} = \int_0^{(1-D)T} dI_L = \int_0^{(1-D)T} \frac{V_o\,dt}{L} = \frac{V_o(1-D)T}{L}$$

As we consider that the converter operates in steady-state conditions, the amount of energy stored in each of its components has to be the same at the beginning and at the end of a commutation cycle. As the energy in an inductor is given by:

$$E = \frac{1}{2}LI_L^2$$

it is obvious that the value of I_L at the end of the Off state must be the same with the value of I_L at the beginning of the On-state, i.e. the sum of the variations of I_L during the on and the off states must be zero:

$$\Delta I_{L_{On}} + \Delta I_{L_{Off}} =$$

Substituting $\Delta I_{L_{On}}$ and $\Delta I_{L_{Off}}$ by their expressions yields:

$$\Delta I_{L_{On}} + \Delta I_{L_{Off}} = \frac{V_i DT}{L} + \frac{V_o(1-D)T}{L} = 0$$

This can be written as:

$$\frac{V_o}{V_i} = \frac{-D}{1-D}$$

This in return yields that:

$$D = \frac{V_o}{V_o - V_i}$$

From the above expression it can be seen that the polarity of the output voltage is always negative (because the duty cycle goes from 0 to 1), and that its absolute value increases with D, theoretically up to minus infinity when D approaches 1. Apart from the polarity, this converter is either step-up (a boost converter) or step-down (a buck converter). Thus it is named a buck–boost converter.

Discontinuous Mode

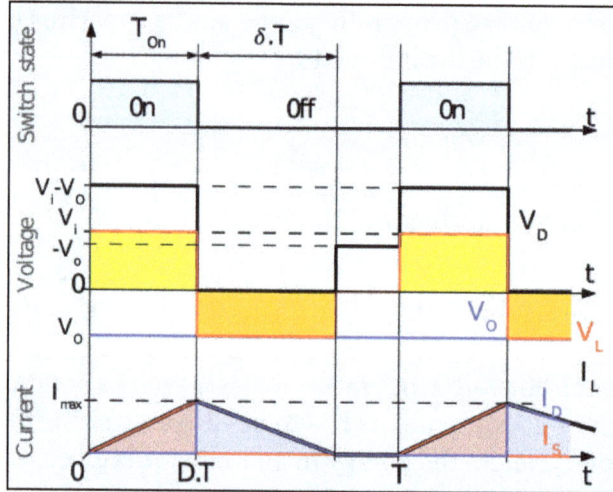

Waveforms of current and voltage in a buck–boost converter operating in discontinuous mode.

In some cases, the amount of energy required by the load is small enough to be transferred in a time smaller than the whole commutation period. In this case, the current through the inductor falls to zero during part of the period. The only difference in the principle described is that the inductor is completely discharged at the end of the commutation cycle. Although slight, the difference has a strong effect on the output voltage equation. It can be calculated as follows:

Because the inductor current at the beginning of the cycle is zero, its maximum value:

$I_{L_{Max}}$ (at $t = DT$) is

$$I_{L_{max}} = \frac{V_i \, DT}{L}$$

During the off-period, I_L falls to zero after $\delta.T$:

$$I_{L_{max}} + \frac{V_o \, \delta T}{L} = 0$$

Using the two previous equations, δ is:

$$\delta = -\frac{V_i \, D}{V_o}$$

The load current I_o is equal to the average diode current (I_D). As can be seen on figure, the diode current is equal to the inductor current during the off-state. Therefore, the output current can be written as:

$$I_o = \overline{I_D} = \frac{I_{L_{max}}}{2}\delta$$

Replacing $I_{L_{max}}$ and δ by their respective expressions yields:

$$I_o = -\frac{V_i DT}{2L}\frac{V_i D}{V_o} = -\frac{V_i^2 D^2 T}{2LV_o}$$

Therefore, the output voltage gain can be written as:

$$\frac{V_o}{V_i} = -\frac{V_i D\ T}{LI_o}$$

Compared to the expression of the output voltage gain for the continuous mode, this expression is much more complicated. Furthermore, in discontinuous operation, the output voltage not only depends on the duty cycle, but also on the inductor value, the input voltage and the output current.

Limit between Continuous and Discontinuous Modes

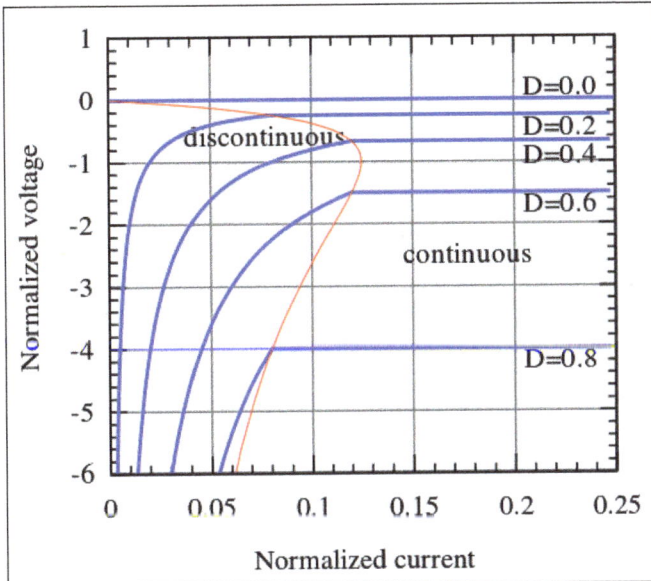

Evolution of the normalized output voltage with the normalized output current in a buck–boost converter.

As told at the beginning, the converter operates in discontinuous mode when low current is drawn by the load, and in continuous mode at higher load current levels. The limit between discontinuous and continuous modes is reached when the inductor current falls to zero exactly at the end of the commutation cycle. with the notations of figure, this corresponds to:

$$DT + \delta T = T$$
$$D + \delta = 1$$

In this case, the output current $I_{o_{\lim}}$ (output current at the limit between continuous and discontinuous modes) is given by:

$$I_{o_{\lim}} = \overline{I_D} = \frac{I_{L_{\max}}}{2}(1-D)$$

Replacing $I_{o_{\lim}}$ by the expression given in the *discontinuous mode* section yields:

$$I_{o_{\lim}} = \overline{I_D} = \frac{I_{L_{\max}}}{2}(1-D)$$

Replacing $I_{\max}$ by the expression given in the discontinuous mode section yields:

$$I_{o_{\lim}} = \frac{V_i DT}{2L}(1-D)$$

As $I_{o_{\lim}}$ is the current at the limit between continuous and discontinuous modes of operations, it satisfies the expressions of both modes. Therefore, using the expression of the output voltage in continuous mode, the previous expression can be written as:

$$I_{o_{\lim}} = \frac{V_i DT}{2L}\frac{V_i}{V_o}(-D)$$

Let's now introduce two more notations:

- The normalized voltage, defined by $|V_o| = \dfrac{V_o}{V_i}$ It corresponds to the gain in voltage of the converter.

- The normalized current, defined by $|I_o| = \dfrac{L}{TV_i}I_o$ The term $\dfrac{TV_i}{L}$ is equal to the maximum increase of the inductor current during a cycle; i.e., the increase of the inductor current with a duty cycle D=1. So, in steady state operation of the converter, this means that $|I_o|$ equals 0 for no output current, and 1 for the maximum current the converter can deliver.

Using these notations, we have:

- In continuous mode $|V_o| = -\dfrac{D}{1-D}$.

- In discontinuous mode, $|V_o| = -\dfrac{D^2}{2|I_o|}.$

- The current at the limit between continuous and discontinuous mode is

$$I_{o_{\lim}} = \frac{V_i T}{2L} D(1-D) = \frac{I_{o_{\lim}}}{2|I_o|} D(1-D)$$ Therefore the locus of the limit between continuous and discontinuous modes is given by $\dfrac{1}{2|I_o|} D(1-D) = 1.$

These expressions have been plotted in figure. The difference in behavior between the continuous and discontinuous modes can be seen clearly.

Principles of Operation of the 4-switch Topology

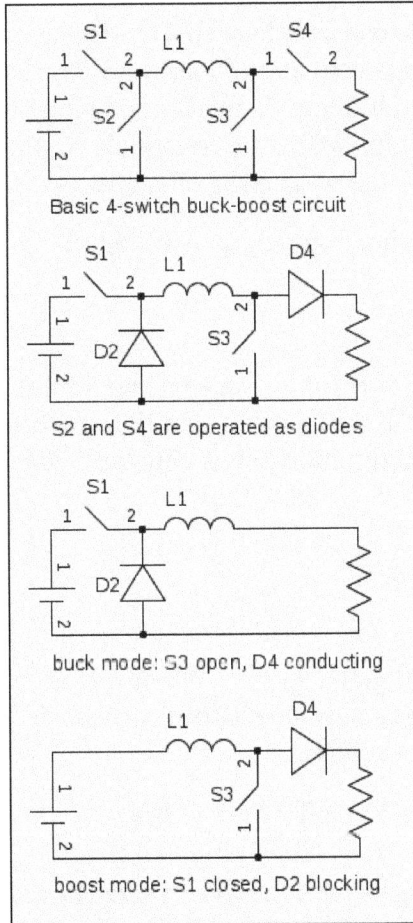

The basics of the 4-switch topology.

The 4-switch converter combines the buck and boost converters. It can operate in either the buck or the boost mode. In either mode, only one switch controls the duty cycle, another is for commutation and must be operated inversely to the former one, and the remaining two switches are in a fixed position. A 2-switch buck-boost converter can be built with two diodes, but upgrading the diodes to FET transistor switches doesn't cost much extra while due to lower voltage drop the efficiency improves.

Non-ideal Circuit

Effect of Parasitic Resistances

In the analysis above, no dissipative elements (resistors) have been considered. That means that the power is transmitted without losses from the input voltage source to the load. However, parasitic resistances exist in all circuits, due to the resistivity of the materials they are made from. Therefore, a fraction of the power managed by the converter is dissipated by these parasitic resistances.

For the sake of simplicity, we consider here that the inductor is the only non-ideal component, and that it is equivalent to an inductor and a resistor in series. This assumption is acceptable because an inductor is made of one long wound piece of wire, so it is likely to exhibit a non-negligible parasitic resistance (R_L). Furthermore, current flows through the inductor both in the on and the off states.

Using the state-space averaging method, we can write:

$$V_i = \overline{V}_L + \overline{V}_S$$

where $\overline{V}_L$ and $\overline{V}_S$ are respectively the average voltage across the inductor and the switch over the commutation cycle. If we consider that the converter operates in steady-state, the average current through the inductor is constant. The average voltage across the inductor is:

$$\overline{V}_L = L\frac{\overline{dI_L}}{dt} + R_L \overline{I}_L = R_L \overline{I}_L$$

When the switch is in the on-state, $V_S = 0$. When it is off, the diode is forward biased (we consider the continuous mode operation), therefore $V_S = V_i - V_o$. Therefore, the average voltage across the switch is:

$$\overline{V}_S = D0 + (1-D)(V_i - V_o) = (1-D)(V_i - V_o)$$

The output current is the opposite of the inductor current during the off-state. the average inductor current is therefore:

$$\overline{I}_L = \frac{-I_o}{1-D}$$

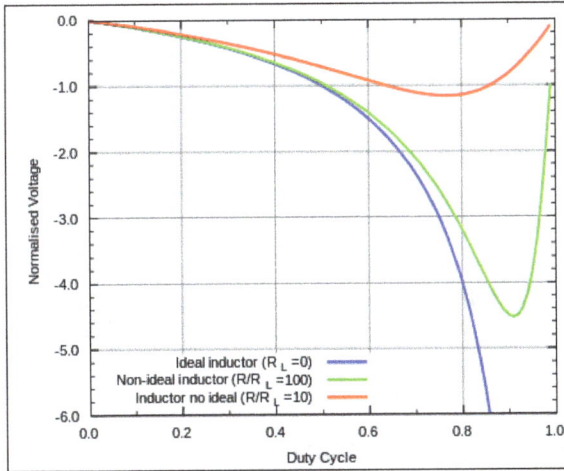

Evolution of the output voltage of a buck–boost converter with the duty cycle when
the parasitic resistance of the inductor increases.

Assuming the output current and voltage have negligible ripple, the load of the converter can be considered purely resistive. If R is the resistance of the load, the above expression becomes:

$$\overline{I}_{\mathrm{L}} = \frac{-V_o}{(1-D)R}$$

Using the previous equations, the input voltage becomes:

$$V_i = R_{\mathrm{L}}\frac{-V_o}{(1-D)R} + (1-D)(V_i - V_o)$$

This can be written as:

$$\frac{V_o}{V_i} = \frac{-D}{\dfrac{R_{\mathrm{L}}}{R(1-D)} + 1 - D}$$

If the inductor resistance is zero, the equation above becomes equal to the one of the *ideal* case. But when R_{L} increases, the voltage gain of the converter decreases compared to the ideal case. Furthermore, the influence of R_{L} increases with the duty cycle.

AC TO AC CONVERTER

A solid-state AC-to-AC converter converts an AC waveform to another AC waveform, where the output voltage and frequency can be set arbitrarily.

DC Link Converters

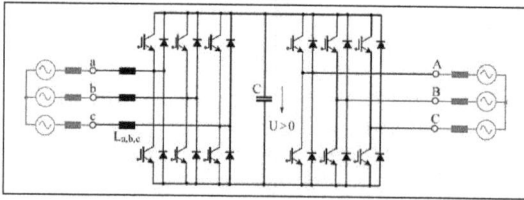

Topology of (regenerative) voltage-source inverter AC/DC-AC converter.

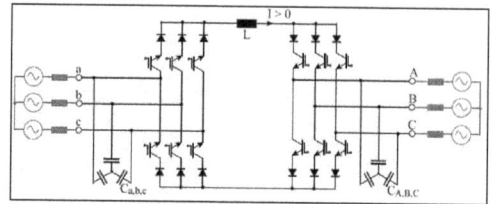

Topology of current-source inverter AC/DC-AC converter.

There are two types of converters with DC link:

- Voltage-source inverter (VSI) converters: In VSI converters, the rectifier consists of a diode-bridge and the DC link consists of a shunt capacitor.

- Current-source inverter (CSI) converters: In CSI converters, the rectifer consists of a phase-controlled switching device bridge and the DC link consists of 1 or 2 series inductors between one or both legs of the connection between rectifier and inverter.

Any dynamic braking operation required for the motor can be realized by means of braking DC chopper and resistor shunt connected across the rectifier. Alternatively, an anti-parallel thyristor bridge must be provided in the rectifier section to feed energy back into the AC line. Such phase-controlled thyristor-based rectifiers however have higher AC line distortion and lower power factor at low load than diode-based rectifiers.

An AC-AC converter with approximately sinusoidal input currents and bidirectional power flow can be realized by coupling a pulse-width modulation (PWM) rectifier and a PWM inverter to the DC-link. The DC-link quantity is then impressed by an energy storage element that is common to both stages, which is a capacitor C for the voltage DC-link or an inductor L for the current DC-link. The PWM rectifier is controlled in a way that a sinusoidal AC line current is drawn, which is in phase or anti-phase (for energy feedback) with the corresponding AC line phase voltage.

Due to the DC-link storage element, there is the advantage that both converter stages are to a large extent decoupled for control purposes. Furthermore, a constant, AC line independent input quantity exists for the PWM inverter stage, which results in high utilization of the converter's power capability. On the other hand, the DC-link energy storage element has a relatively large physical volume, and when electrolytic capacitors are used, in the case of a voltage DC-link, there is potentially a reduced system lifetime.

Cycloconverters

A cycloconverter (CCV) or a cycloinverter converts a constant voltage, constant frequency AC waveform to another AC waveform of a lower frequency by synthesizing the output waveform from segments of the AC supply without an intermediate DC link. There are two main types of CCVs, circulating current type or blocking mode type, most commercial high power products being of the blocking mode type.

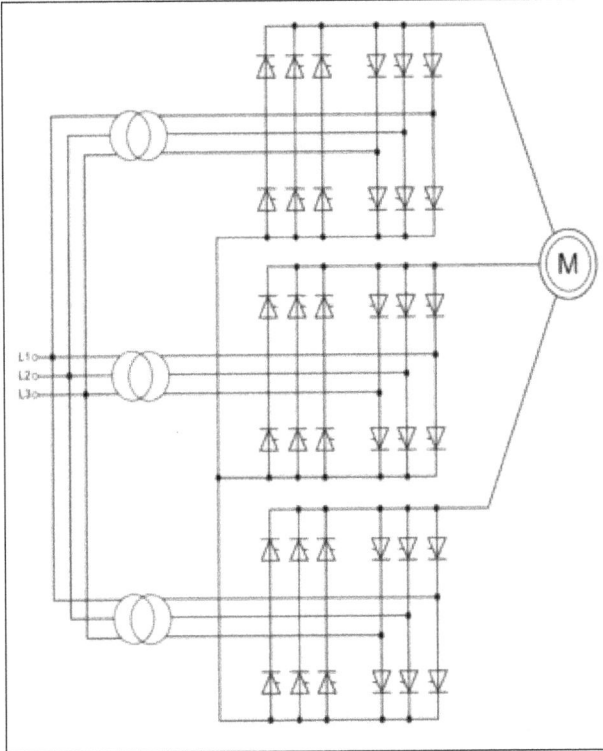

Topology of blocking mode cycloconverter.

Characteristics

Whereas phase-controlled SCR switching devices can be used throughout the range of CCVs, low cost, low-power TRIAC-based CCVs are inherently reserved for resistive load applications. The amplitude and frequency of converters' output voltage are both variable. The output to input frequency ratio of a three-phase CCV must be less than about one-third for circulating current mode CCVs or one-half for blocking mode CCVs Output waveform quality improves as the *pulse number* of switching-device bridges in phase-shifted configuration increases in CCV's input. In general, CCVs can be with 1-phase/1-phase, 3-phase/1-phase and 3-phase/3-phase input/output configurations, most applications however being 3-phase/3-phase.

Applications

The competitive power rating span of standardized CCVs ranges from few megawatts up to many tens of megawatts. CCVs are used for driving mine hoists, rolling mill main motors, ball mills for ore processing, cement kilns, ship propulsion systems, slip power recovery wound-rotor induction motors (i.e., Scherbius drives) and aircraft 400 Hz power generation. The variable-frequency output of a cycloconverter can be reduced essentially to zero. This means that very large motors can be started on full load at very slow revolutions, and brought gradually up to full speed. This is invaluable with, for example, ball mills, allowing starting with a full load rather than the alternative of having to start the mill with an empty barrel then progressively load it to full capacity. A fully loaded "hard start" for such equipment would essentially be applying full power to a stalled motor. Variable speed and reversing are essential to processes such as hot-rolling steel mills. Previously, SCR-controlled DC motors were used, needing regular brush/commutator servicing and delivering lower efficiency. Cycloconverter-driven synchronous motors need less maintenance and give greater reliability and efficiency. Single-phase bridge CCVs have also been used extensively in electric traction applications to for example produce 25 Hz power in the U.S. and 16 2/3 Hz power in Europe.

Whereas phase-controlled converters including CCVs are gradually being replaced by faster PWM self-controlled converters based on IGBT, GTO, IGCT and other switching devices, these older classical converters are still used at the higher end of the power rating range of these applications.

Harmonics

CCV operation creates current and voltage harmonics on the CCV's input and output. AC line harmonics are created on CCV's input accordance to the equation, whereas phase-controlled converters including CCVs are gradually being replaced by faster PWM self-controlled converters based on IGBT, GTO, IGCT and other switching devices, these older classical converters are still used at the higher end of the power rating range of these applications.

$$f_h = f_1(kq\pm1) \pm 6nf_o$$

where,

f_h = harmonic frequency imposed on the AC line,

k and n = integers,

q = pulse number (6, 12 . . .),

f_o = output frequency of the CCV.

- Equation's 1st term represents the *pulse number* converter harmonic components starting with six-pulse configuration,

- Equation's 2nd term denotes the converter's sideband characteristic frequencies including associated interharmonics and subharmonics.

Matrix Converter

A matrix converter is defined as a converter with a single stage of conversion. It utilizes bidirectional controlled switch to achieve automatic conversion of power from AC to AC. It provides an alternative to PWM voltage rectifier (double sided).

Matrix converters are characterized by sinusoidal waveforms that show the input and output switching frequencies. The bidirectional switches make it possible to have a controllable power factor input. In addition, the lack of DC links ensures it has a compact design. The downside to matrix converters is that they lack bilateral switches that are fully controlled and able to operate at high frequencies. Its voltage ratio that is output to input voltage is limited.

There are three methods of matrix converter control:

- Space vector modulation.

- Pulse width modulation.

- Venturi - analysis of function transfer.

The main advantage of matrix converter is elimination of dc link filter. Zero switching loss devices can transfer input power to output power without any power loss. But practically it does not exist. The switching frequency of the device decides the THD of the converter. Maximum power transfer to the load is decided by nature of the control algorithm. Matrix converter has a maximum input output voltage transfer ratio limited to 87 % for sinusoidal input and output waveforms, which can be improved. Further, matrix converter requires more semiconductor devices than a conventional AC-AC indirect power frequency converter. Since monolithic bi-directional switches are available they are used for switching purpose.

Matrix converter is particularly sensitive to the disturbances of the input voltage to the system,. This can be attenuated by intelligent control technique and the fuzzy controller has a least effect due to input side disturbance. The simulation of single phase matrix converter and three phase matrix converter is obtained from a simplified simulation model. Here single phase matrix converter, types of switching patterns, and simulation model of the matrix converter are described.

Single Phase Matrix Converter

The AC/AC converter is commonly classified as an indirect converter which utilizes a dc link between the two ac systems and converter that provides direct conversion. This converter consists of two converter stages and energy storage element, which convert input ac to dc and then reconverting dc back to output ac with variable amplitude and frequency.

The operation of this converter stages is decoupled on an instantaneous basis by the energy storage elements and controlled independently, so long as the average energy flow is equal. Figure shows the single phase matrix converter switching arrangement.

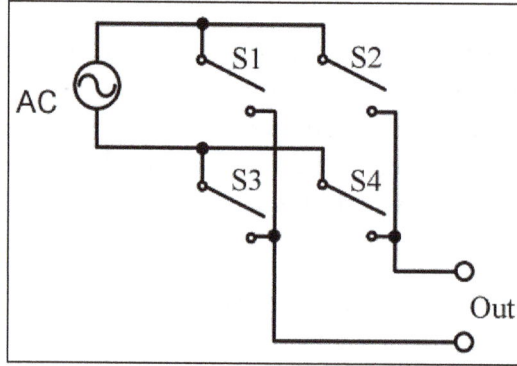

Single Phase Matrix Converter.

Figures show the operation of the single matrix converter in four modes of operation. Figure shows that generation of pulse width modulation gate signal using MATLAB. This may be implemented using logic gates. In this the pulse generation is obtained by logic gates. In figure 1A, 1B, 2A and 2B are gate drive pulse to the single phase matrix converter.

The matrix converter requires a bidirectional switch capable of blocking voltage and conducting current in both directions. Unfortunately there is no discrete component that fulfils these needs. To overcome this problem the common emitter anti-parallel IGBT , diode pair is used. Diodes are in place to provide reverse blocking capability to the switch module.

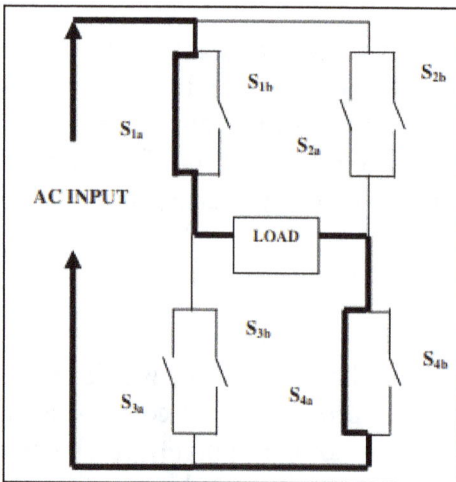

S1a and S4a Switched on during
Positive Half Cycle.

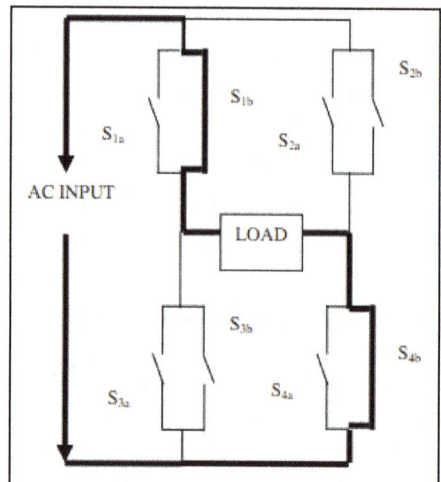

S1b and S4b Switched on during
Negative Half Cycle.

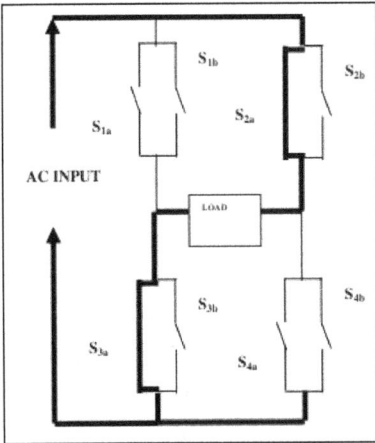

S2a and S3a Switched on during
Positive Half Cycle.

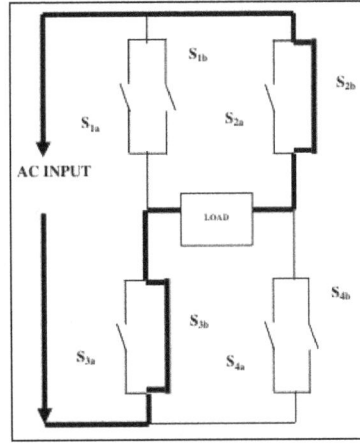

S2b and S3b Switched on during
Negative Half Cycle.

For an example consider the development of a pulse width modulation technique using Matlab Simulink. Figure represents development of sinusoidal pulse width modulation gate signal using Matlab logic blocks. Sinusoidal pulse width modulation gate pulse can be generated using triangular signal. Triangular signal and sinusoidal reference signal are compared to get the sinusoidal pulse width modulation gate pulse. The frequency of the sinusoidal function is fixed. Variable pulse width of the gate pulse can be obtained by using triangular signal of varying time period. Modulation index gives the information about "ON" time of the gate pulse. It is calculated as switch on time divided by switch ON time plus OFF time of the device. Here 0.7 modulation index is taken for the simulation.

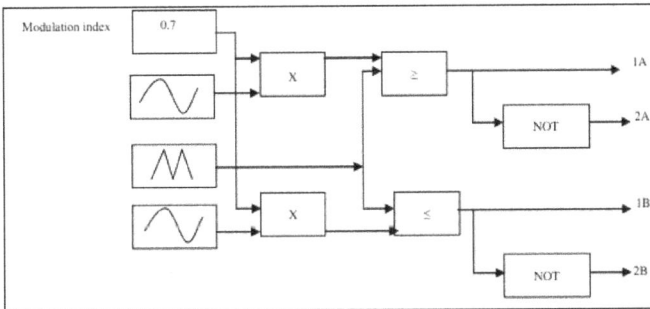

Generation of Sinusoidal Pulse Width Modulation Gate Pulse.

Three Phase Matrix Converter

The instantaneous power flow does not have to equal power output. The difference between the input and output power must be absorbed or delivered by an energy storage element within the converter.

The matrix converter replaces the multiple conversion stages and the intermediate energy storage element by a single power conversion stage, and uses a matrix of semiconductor bidirectional switches connecting input and output terminals. With this general arrangement of switches, the power flow through the converter can reverse. Because of the absence of any energy storage element, the instantaneous power input must be equal to the power output, assuming idealized zero-loss switches.

However, the reactive power input does not have to equal power output. It can be said again that the phase angle between the voltages and currents at the input can be controlled and does not have to be the same as at the output.

Three phase matrix converter consists of nine bidirectional switches. It has been arranged into three groups of three switches. Each group is connected to each phase of the output. These arrangements of switches can connect any input phase. In the figure filled circle shows a closed switch. These 3 X 3 arrangements can have 512 switching states. Among them only 27 switching states are permitted to operate this converter. For safe operation, it should follow the given rules:

- Do not connect two different input lines to the same output line(input short circuited).

- Do not disconnect the output line circuits (output open circuited).

Figures are showing different operating states of matrix converter. Here A, B and C are input phase voltage connected to the output phase. Figure shows synchronous operating state vectors of three matrix converter. It shows that the converter switches are switched on rotational basis. In this case no two switches in a leg are switched on simultaneously. These states will not generate gate pulse when one phase of the supply is switched off.

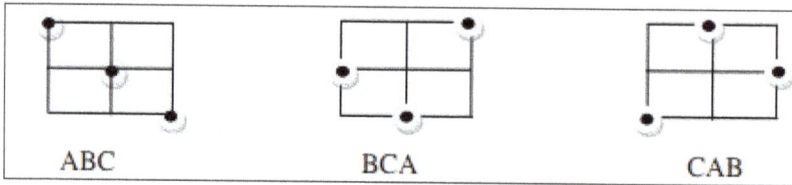

| ABC | BCA | CAB |

Matrix Converter Rotating Vectors (Synchronous Vectors).

Figure shows inverse operating state vectors of three matrix converter. In this any one phase is rotated in such a way that it connects all the output phase in a cycle of operation. This operation may be selected during reverse operation of induction motor.

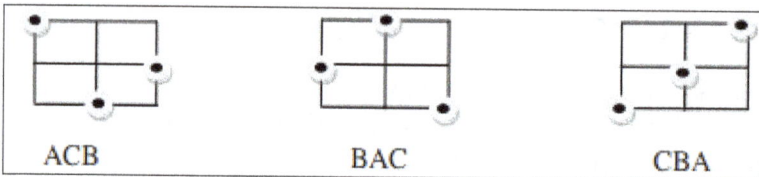

| ACB | BAC | CBA |

Matrix Converter Rotating Vectors (Inverse Operation).

Figure shows zero vector of the matrix converter. Here all the output phases are connected in a single input line. It leads to damage to the device. Because three phase loads are directly connected to the single phase line.

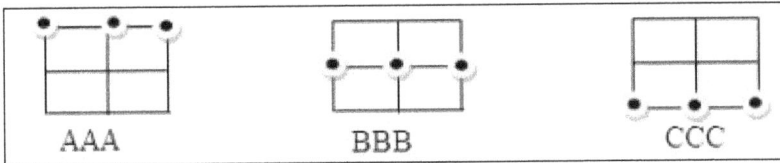

Matrix Converter Zero Vectors.

Figure shows active vectors of the matrix converter which are the operating states in direct conversion. There are 18 operating states are available. We can select any combination for the operation of matrix converter.

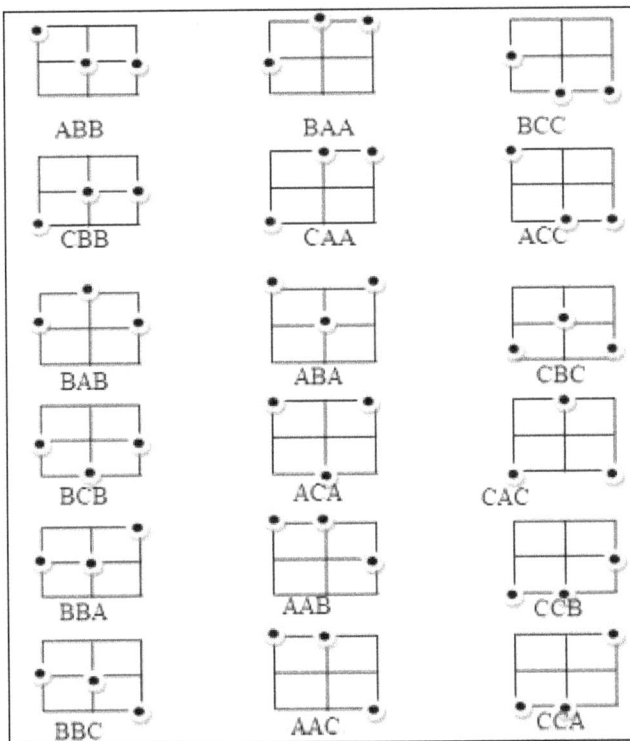

Matrix Converter Active Vectors (Pulsating).

Carrier based PWM

Direct AC to AC converter is used in variable speed electric drive for variable voltage and variable frequency source. The important task of SVPWM for the matrix converters is to generate the required output voltages while controlling the input currents or the power factor as required. One limitation of the matrix converter is that the maximum output voltage available is limited to 86.6 % of the input voltage in the linear modulation range.

There are basically three PWM schemes for matrix converter control. These include carrier based PWM, space vector PWM and selective harmonics elimination PWM methods. In case of inverter the carrier based PWM methods can be advantageously utilized in: 1) controlling common mode voltage and 2) controlling complicated inverter topologies like multi level inverter. The space vector modulation method presents a good tool for a balanced three phase input. However, its algorithm for implementation is rather difficult, particularly if common mode control is applied to improve PWM performance. The switching frequency of the matrix converter was derived from the two switching functions obtained from the PWM converter and the inverter modulation. While many calculations are required, this method has the advantage that the well established space vector PWM method of VSI can be applied to the matrix converter modulation. Input power factor control, maximum voltage utilization and the modulation under unbalanced input voltage have been possible in carrier based PWM. The SVPWM has flexibility and many good points as a common industrial practice. But, it needs more calculations and tables for switching pattern in accordance with the input current and output voltage sectors. Implementation is unintuitive because the gating signals are made from the duty cycles of the effective space vectors which are calculated by the equations.

Important function of the space vector modulation is to determine the pulse width for active vectors within each sampling interval which contribute fundamental components in the line to line voltages. The optimal sequence of the pulse within the sampling interval leads to a superior performing space vector modulation pulse.

The SVPWM may be classified as direct or indirect modulation according to the structure of their modulation matrices. The direct modulation views the modulation matrix as a direct AC-AC conversion, while the indirect modulation assumes a virtual DC link and decomposes the modulation matrix into the rectifier and inverter modulation matrices. Otherwise, it is also possible to categorize a PWM method as a scalar or space vector modulation depending on the three phase voltages and currents.

Generally indirect space vector PWM needs a look-up table to select the necessary active and zero vectors depending on the status of the voltage and current vectors. The switching times of the selected space vectors are obtained through the combination of the duty cycles of the rectifier and the inverter. The switching signals of the matrix switches are determined from switching states and switching times of the selected space vectors. Therefore, implementation of the indirect space vector modulation requires complicated duty cycle calculations and look-up table to determine the PWM switching signals. The calculated duty cycles are also not straightforwardly related to the instantaneous phase values. Furthermore, it is not known yet whether indirect space vector PWM can be realized as a carrier based PWM. Consequently, the space vector modulation for matrix converters is more complex than that of two level inverters. The direct space vector modulation methods usually have 27 space vectors of the matrix converter. From these descriptions, it is obvious that SVPWM is computationally much more demanding approach than carrier based PWM.

A carrier based PWM modulator combines modulation signal and carrier signal. Therefore in each carrier signal period each output of the converter legs is switching between the positive or negative rail of the dc link. If the reference signal is greater than the carrier signal, then the active device corresponding to that carrier is switched on; and if the reference signal is less than the carrier signal, then the active device corresponding to that carrier is switched off. Figure shows the development of carrier - based PWM technique. Here $V_a{}^*$, $V_b{}^*$ and $V_c{}^*$ are 3 phase AC input voltage. S1, S2, S3, S4, S5 and S6 are switches of 3 phase voltage source inverter.

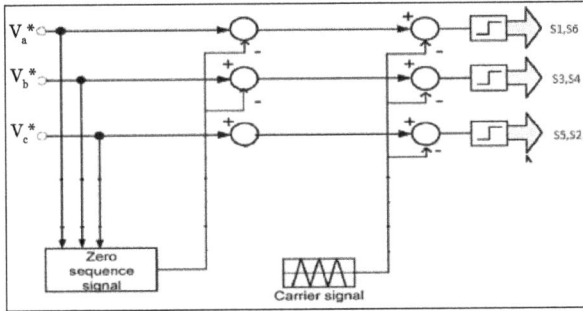

Carrier - based PWM for 3 phase VSI.

The important performance of a carrier-based PWM modulator is found by its modulation signals. However, a carrier affects the superior performance of the modulator too. The PWM signal is generated by comparing a sinusoidal modulating signal with a triangular signal having double edge or a saw tooth signal having single edge carrier signal. The frequency of the carrier is normally kept much higher compared to that of the modulating signal.

SVPWM is the most popular one due to its simplicity both in hardware and software, and its relatively good performance at low modulation ratio. But the SVPWM becomes very difficult to achieve when the levels of the converter increase. Generally, carrier - based PWM of multilevel inverter can only select four switching states at most, but SVPWM can select more. In general, selection of switching states has more freedom in the Space Vector PWM than the carrier based PWM. Generally, in carrier-based PWM mode the modulated output voltage is smooth and it contains distortion. This carrier-based PWM technique can be advantageous if there are a large number of levels and the levels are taken care of in multilevel inverters. In case of matrix converter which has fixed number of switches Space Vector Pulse Width Modulation is preferred. The carrier based PWM method with the smallest common mode voltage presents a preferable PWM solution for high power and high number level inverters. Then these SVPWM switches should be selected in a proper manner. The matrix converter is a nonlinear controller because it uses nonlinear components. Fuzzy controller is an approach to overcome the waveform quality problem.

The poor power quality can degrade or damage the matrix converter. Improving power quality may be achieved using a three phase series active filter. The active filter may correct the voltage unbalances and regulate it to the desired level. The fuzzy logic controller is used to correct and regulate the unbalance voltage in three phase system matrix converter.

The application of fuzzy logic control seems to be very well suited for controlling such a system. Fuzzy logic deals with problems where the relationship between the inputs and the outputs of the system cannot be expressed mathematically in an easy way but it can be expressed by means of linguistic terms. If–Then rules are used to represent fuzzy inference. The set of If–Then rules defines a process of mapping from a given input voltage to an output voltage using fuzzy logic. This input–output transformation can be represented as an inference table or matrix. Processing an If–Then rule involves evaluating the antecedent, which involves applying fuzzy operators and applying the results to the output. Finally, aggregation of all outputs is needed to form a single fuzzy set. This is done by a process called composition. The most common composition method is the maximum composition, which consists of taking the maximum value of all output fuzzy sets.

This fuzzy logic controller finds the rule that depends upon the error signal. For induction motor application, the fuzzy logic controller decides the new space vector voltage angle that gives the best stator flux and torque response according the voltage vector components. It is possible to reduce the switching frequency in the inverter or matrix converter and diminish the harmonic contents of the stator current signal. It is possible to improve the dynamic response of the stator flux but the torque response has no significant difference. The purpose is to decrease the speed and torque fluctuations of the control system and improve the voltage of the system. Thus it can provide better waveform quality to the operate induction motor.

Commutation Methods in Matrix Converter

The commutation has to be actively controlled at all times. It is important that no two bidirectional switches are switched on at the same instant. This results in short circuit at capacitor input and open circuit at inductive load.

Dead Time Commutation

This type of commutation method is used in the inverter side. It means that load current freewheel to throw antiparallel diode during the dead time period. In case of the matrix converter dead time commutation method is useless. It results in the open circuit at the load side. Then forced spike occurs across the switches. To avoid this snubber clamping devices are provided.

This is a path to the load current during the dead time and hence the design of snubber circuit is difficult.

Current Commutation based on Multiple Steps

This type of commutation uses bidirectional switches. These are reliable in current commutation and obey the basic rules. It can be able to control the direction of the current. This strategy is essential in case of controlled current flow. This commutation technique relies on knowledge of the output current direction. This current direction can be difficult to reliably determine and allow current levels in high power drives. To avoid this problem a technique of using the voltage across the bi-directional switch to determine the current direction has been developed. This technique provides reliable current commutation using an intelligent gate drive circuit which controls the firing of the IGBTs and detects the direction of current flow within the bidirectional switch cell. The current direction information calculated by the active gate drive is passed to all the other gate drivers on the same output leg. In this way all the gate drivers contribute to operate a safe commutation. In matrix converter commutation issue is taken care by matlab simulation. Forced commutation is a employed throughout the process.

References

- U.S. Patent 4257087.: «DC-to-DC switching converter with zero input and output current ripple and integrated magnetics circuits», filed 2 Apr 1979, Retrieved 15 Jan 2017

- Power-electronic-converters: elprocus.com, Retrieved 3 March, 2019

- Stephen Sangwine (2 March 2007). Electronic Components and Technology, Third Edition. CRC Press. P. 73. ISBN 978-1-4200-0768-8

- Power-electronics-matrix-converter, power-electronics: tutorialspoint.com, Retrieved 4 April, 2019

- Iqbal, Sajid; et al. (2014). Study of bifurcation and chaos in dc-dc boost converter using discrete-time map. IEEE International Conference on Mechatronics and Control (ICMC'2014) 2014. Doi:10.1109/ICMC.2014.7231874

- Chierchie, F. Paolini, E.E. Discrete-time modeling and control of a synchronous buck converter .Argentine School of Micro-Nanoelectronics, Technology and Applications, 2009. EAMTA 2009.1–2 October 2009, pp. 5 – 10. ISBN 978-1-4244-4835-7

Permissions

Index

www.ingramcontent.com/pod-product-compliance
Lightning Source LLC
Chambersburg PA
CBHW061936190326

41458CB00009B/2757